Quick Guide

WALLS & CEILINGS

Manufactured in the United States of America

Creative Director: Warren Ramezzana
Editor: Arnold Edelstein
Project Editor: Kimberly Kerrigone
Graphic Designer: Annie Jeon
Production Assistant: Mindy Circelli
Illustrators: James Randolph, Norman Nuding
Technical Reviewer: Jim Barrett

Cover Design: Warren Ramezzana
Cover Illustrations: Moffit Cecil

Electronic Prepress: M. E. Aslett Corporation
Printed at: Quebecor Printing Inc.

Current Printing (last digit)
10 9 8 7 6 5 4 3

Quick Guide: Walls & Ceilings
LC: 92-81622
ISBN: 1-880029-08-1 (paper)

CREATIVE HOMEOWNER PRESS®
A DIVISION OF FEDERAL MARKETING CORP.
24 PARK WAY
UPPER SADDLE RIVER, NJ 07458

C O N T E N T S

Walls 5

Ceilings 53

Glossary 79

Index 80

SAFETY FIRST

Though all the designs and methods in this book have been tested for safety, it is not possible to overstate the importance of using the safest construction methods possible. What follows are reminders; some do's and don'ts of basic carpentry. They are not substitutes for your own common sense.

- *Always* use caution, care, and good judgment when following the procedures described in this book.
- *Always* be sure that the electrical setup is safe; be sure that no circuit is overloaded, and that all power tools and electrical outlets are properly grounded. Do not use power tools in wet locations.
- *Always* read container labels on paints, solvents, and other products; provide ventilation, and observe all other warnings.
- *Always* read the tool manufacturer's instructions for using a tool, especially the warnings.
- *Always* use holders or pushers to work pieces shorter than 3 inches on a table saw or jointer. Avoid working short pieces if you can.
- *Always* know the limitations of your tools. Do not try to force them to do what they were not designed to do.
- *Always* make sure that any adjustment is locked before proceeding. For example, check the rip fence on a table saw or the bevel adjustment on a portable saw before starting to work.
- *Always* clamp small pieces firmly to a bench or other work surfaces when sawing or drilling.
- *Always* wear the appropriate rubber or work gloves when handling chemicals, heavy construction or when sanding.
- *Always* have a plaster sample tested for asbestos content before removing old plaster walls. If asbestos is found, contact a licensed abatement contractor regarding the removal and disposal of hazardous building materials.
- *Always* wear a disposable mask when working with odors, dusts or mists. Use a special respirator when working with toxic substances.
- *Always* wear eye protection, especially when using power tools or striking metal on metal or concrete; a chip can fly off, for example, when chiseling concrete.
- *Always* be aware that there is never time for your body's reflexes to save you from injury from a power tool in a dangerous situation; everything happens too fast. Be *alert!*
- *Always* keep your hands away from the business ends of blades, cutters and bits.
- *Always* hold a portable circular saw with both hands so that you will know where your hands are.
- *Always* use a drill with an auxiliary handle to control the torque when large size bits are used.
- *Always* check your local building codes when planning new construction. The codes are intended to protect public safety and should be observed to the letter.
- *Never* work with power tools when you are tired or under the influence of alcohol or drugs.
- *Never* cut very small pieces of wood or pipe. Whenever possible, cut small pieces off larger pieces.
- *Never* change a blade or a bit unless the power cord is unplugged. Do not depend on the switch being off; you might accidentally hit it.
- *Never* work in insufficient lighting.
- *Never* work while wearing loose clothing, hanging hair, open cuffs, or jewelry.
- *Never* work with dull tools. Have them sharpened, or learn how to sharpen them yourself.
- *Never* use a power tool on a workpiece that is not firmly supported or clamped.
- *Never* saw a workpiece that spans a large distance between horses without close support on either side of the kerf; the piece can bend, closing the kerf and jamming the blade, causing saw kickback.
- *Never* support a workpiece with your leg or other part of your body when sawing.
- *Never* carry sharp or pointed tools, such as utility knives, awls, or chisels in your pocket. If you want to carry tools, use a special-purpose tool belt with leather pockets and holders.

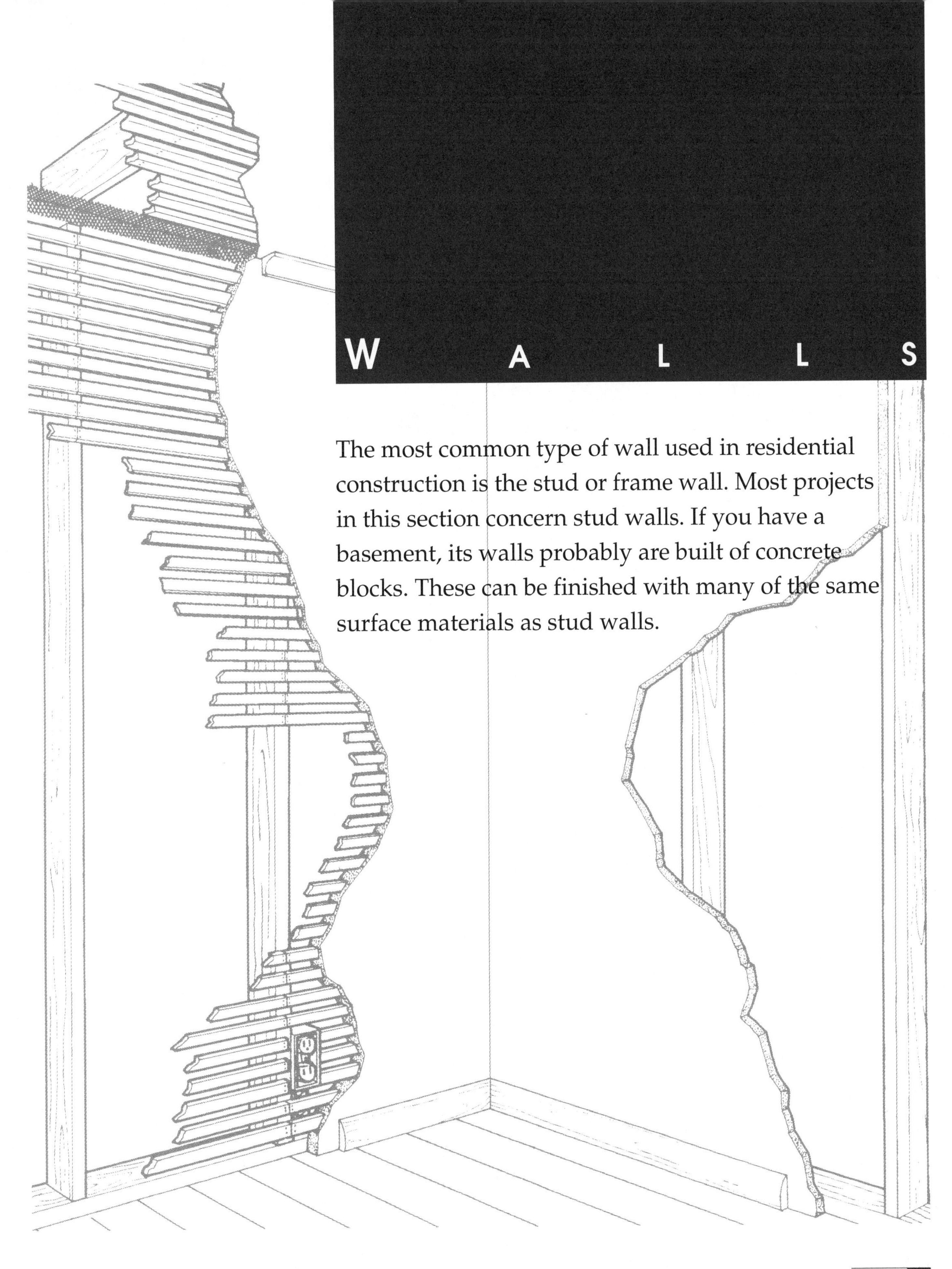

W A L L S

The most common type of wall used in residential construction is the stud or frame wall. Most projects in this section concern stud walls. If you have a basement, its walls probably are built of concrete blocks. These can be finished with many of the same surface materials as stud walls.

Anatomy of Walls

Structure

A stud wall consists of vertical members called studs attached at the top and bottom to horizontal members called plates. The plates in a wall are attached to the floor and ceiling.
At corners and where adjacent walls meet, walls are attached along studs. In most houses these studs and plates are lumber, usually 2x4s or 2x3s in interior walls; in some newer construction, lightweight aluminum studs and plates have replaced wood. The framework of studs may provide space for wiring and outlets, plumbing, duct work or insulation. The main job of the framework is to support a surface of some kind, usually either plaster or wallboard. This surface may in turn be covered by another, such as paneling or tile.

Interior walls are separated into two categories by the function they serve in the structure of the house: Non-bearing walls are simply partitions, dividing open space into rooms, while bearing walls hold up the roof or the floor above.

Surface

Until around 1930, most homes were built with walls made of wood lath plaster. When plastering a wall, the lath is nailed to studs and joists in staggered groups of four to six. Gaps between strips allow two to three coats of plaster to settle into the 1/4-inch spaces between laths, hardening and forming tight grips. Metal lath, which is either the expanded metal or the diamond-shaped type, is used to reinforce the interior corners of the room.

Most walls in houses built since the 1940s are surfaced with wallboard, also known as drywall, or by its trade name, Sheetrock. Strong, inexpensive, and easy to install, wallboard has a core of gypsum. It is covered with a paper surface which may be used as the wall surface, or serve as a backing for another surface material.

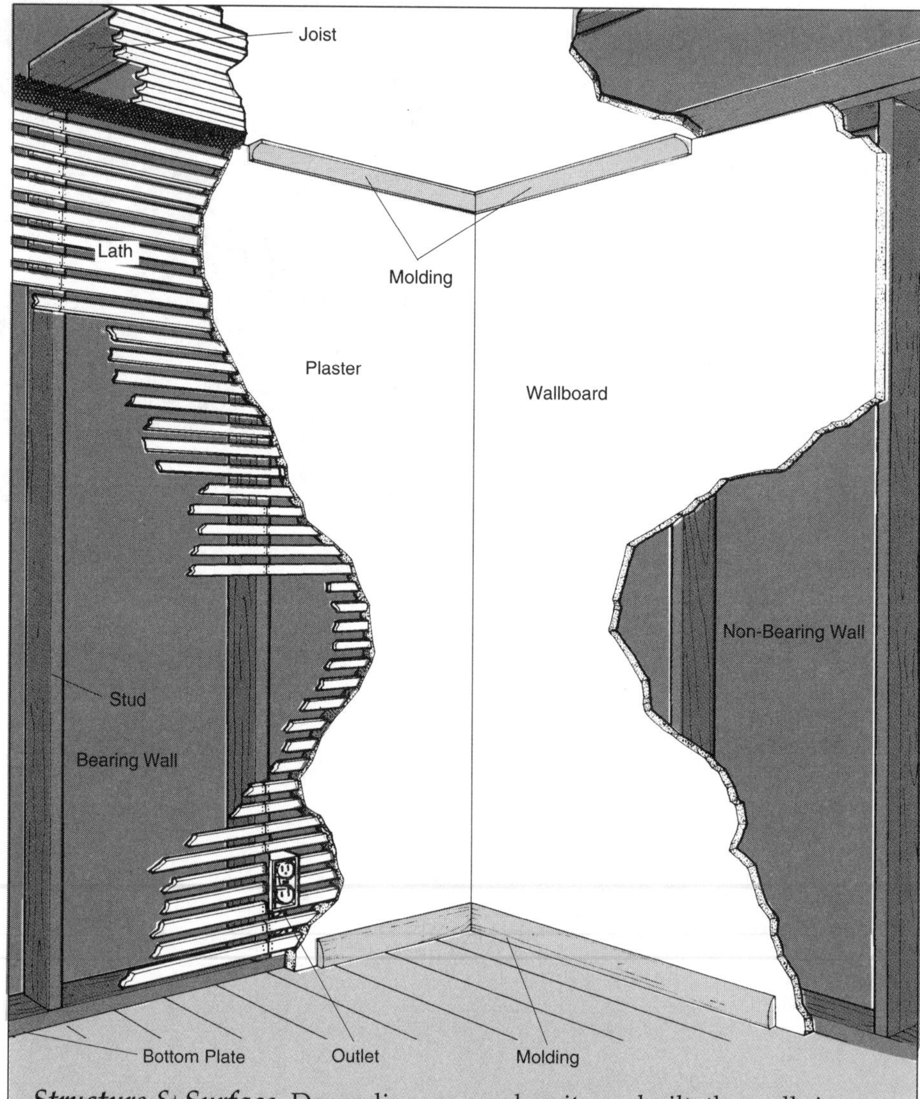

Structure & Surface. Depending upon when it was built, the walls in your home are made of plaster or wallboard. Both provide smooth, sturdy surfaces.

Standard wallboard is available in 4x8-foot panels that are 3/8 or 1/2 inch thick. It is fastened to studs or joists with nails or screws and sometimes adhesive as well. The seams between panels are sealed with paper tape and joint compound.

Waterproof (green) wallboard is available for walls subject to damp or wet conditions, such as in a bathroom.

Sometimes you can completely change the appearance of existing walls by merely bonding a new surface to the old. However, a wall may be too badly deteriorated, to take a new surface without some modification.

Caution

Always have a plaster sample tested for asbestos content before removing old plaster walls.
If asbestos is found, contact a licensed abatement contractor regarding the removal and disposal of hazardous building materials.

Typically the removal of asbestos requires not only a special vacuum with filters that remove dust from surfaces, but special "air scrubbers" that remove dust from the air as well. The person doing the work must wear a respirator which is connected to an outside air supply.

Repairing Wallboard & Plaster

Walls are subject to a variety of assaults that can damage their surface, especially in a house with children. Besides the everyday wear and tear of impacts and gouges, walls also can develop cracks as a house settles. This section covers repairs to damaged wallboard and the two repairs most often needed for plaster. Large repairs to plaster are beyond the scope of this book.

If you have learned the art of taping and feathering wallboard, repairing it will present few problems. Compared with plaster walls, wallboard is relatively easy to repair because damaged areas can be removed and replaced, but plaster must be stripped down to the lath beneath, and then built up again.

The most common problem with wallboard is that as the house settles, seams open and nails pop up and show through the wallcovering. Repairing a popped nail is a simple job: Drive the nail back into the stud, dimpling the surface as shown on page 28. To add holding power, drive another wallboard nail into the stud about 2 inches above or below the one that popped. Then plaster over the new dimples as shown on page 30. Seams that have opened up should be taped and plastered again as shown on pages 30-31. Wallboard repairs require the same tools as wallboard installation and finishing.

Repairing small cracks and holes in a plaster surface is not much different from working with joint compound on wallboard. You will need patching plaster (sold as a powder to be mixed at home) for the base of the new surface and joint compound or vinyl spackling for the finish. The ideal time to make repairs to wall surfaces is when you are about to repaint a room. If you are not refinishing the whole wall under repair, prime the repair and paint or patch it with wallpaper to match the rest of the wall.

Repairing Wallboard

1 Cutting Out the Damage. Wallboard breaks more easily than a plaster wall. A hard knock, such as from a doorknob flung against it, will cause either a noticeable dent or a hole. To replace a piece of wallboard that is damaged, draw a rectangle around it using a carpenter's square to keep the edges straight and the corners at a 90-degree angle. This will make it easier to get a good fit with the replacement piece. Drill starter holes inside opposite corners, cut the piece with a keyhole saw, and pull it out. If the damaged area is large, cut back to the nearest studs on each side.

2 Adding New Bracing. Cut two pieces of 1x3, each about 6 inches longer than the vertical sides of the hole, as braces for the patch. Insert a brace in the opening and hold it vertically against one edge, centered, so that half the width of the brace is behind the wall and half is showing through the opening. Attach above and below the opening with wallboard screws through the wall into the brace; drive additional screws into the side of the brace in the middle or at 6-inch intervals if the opening is larger than 8 inches. Repeat with the second brace on the other side of the opening. Do not use nails—they will break the wallboard. Cut a wallboard patch the size of the hole, fit it in place and attach with screws through the patch into the braces. If you have cut back to studs, nail strips of 1x2 flush against their edges to provide a nailing surface for the patch.

3 Taping and Feathering the Patch. Finish the seams around the patch with tape and joint compound (see pages 30-31). Feathering, applying successively wider layers of compound on each side of tape, helps blend the seam into bare wallboard.

1 Drill a starter hole in the corners of the outline and cut out the damaged piece. Leave a neat hole to fill.

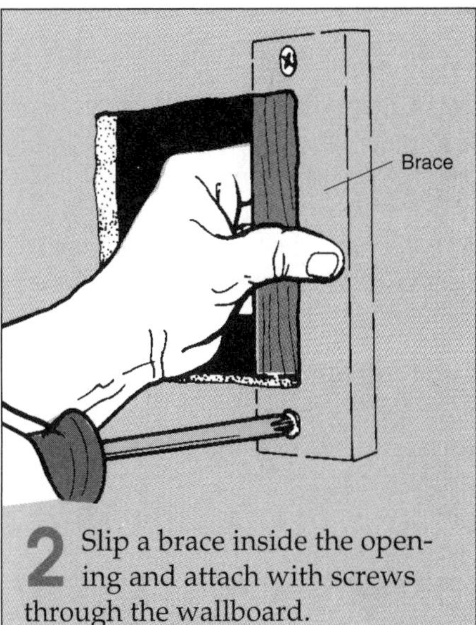

2 Slip a brace inside the opening and attach with screws through the wallboard.

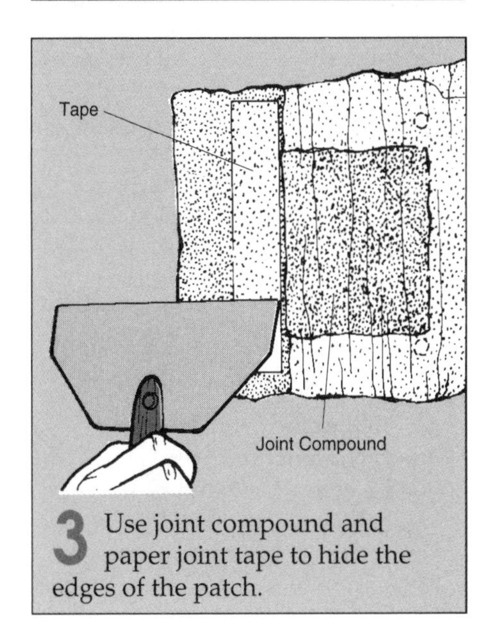

3 Use joint compound and paper joint tape to hide the edges of the patch.

Repairing a Hole with Patching Plaster

1 Inserting the Screen. Fix an 18-inch length of string through a piece of ordinary wire screen cut slightly larger than the hole to be repaired. Clear any loose gypsum and wallboard paper from the edge of the hole. Wet the edge on the inside and coat liberally with patching plaster, then put the screen through the hole and pull it flat against the new plaster.

2 Plastering Over the Screen. Tie the screen to a dowel, pencil or other similar anchor to hold it. Plaster the screen not quite flush with the wall, then tighten the string slightly by twisting the anchor. Allow to dry for a few hours.

3 Cutting the String. After the plaster is well set, cut the string at the screen. Moisten the plaster around the edge of the remaining hole and fill with plaster. Apply a second coat over the entire patch to make it almost flush with the wall and let it dry.

4 Finishing the Patch. Use joint compound to finish the repair, covering the hole and feathering the edges as in taping wallboard joints. Let dry for a day, then sand smooth.

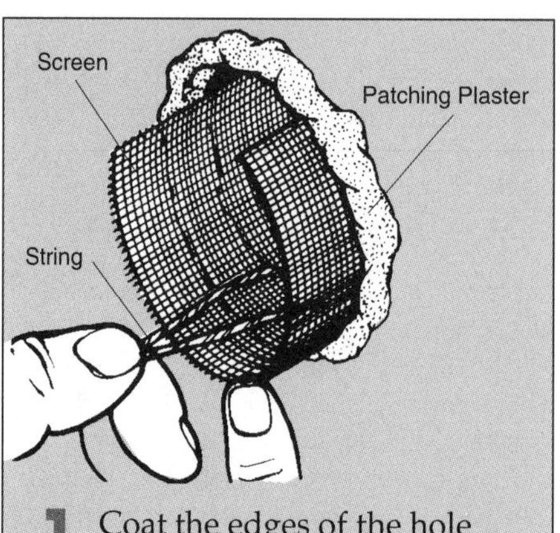

1 Coat the edges of the hole with patching plaster before inserting the screen so it will be fixed to the wall.

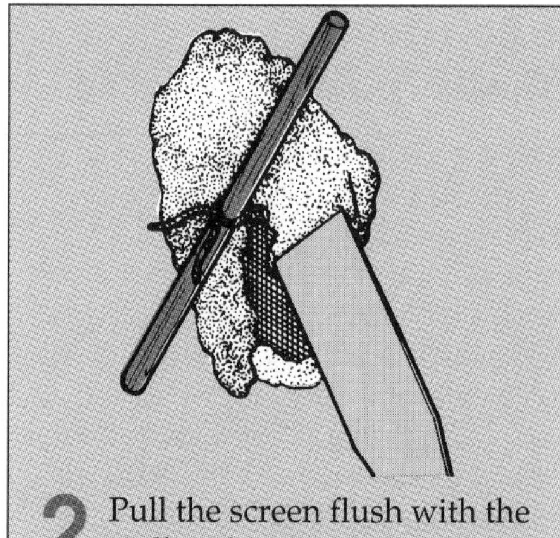

2 Pull the screen flush with the wall, tighten it, and cover with a layer of patching plaster.

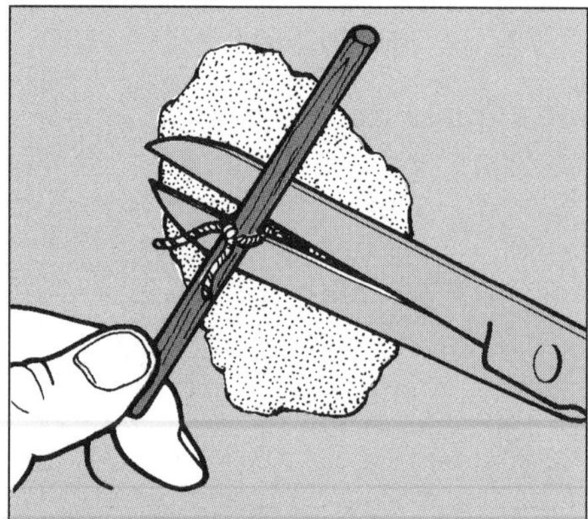

3 When the first layer of plaster is dry, cut away the string and apply a second layer of plaster.

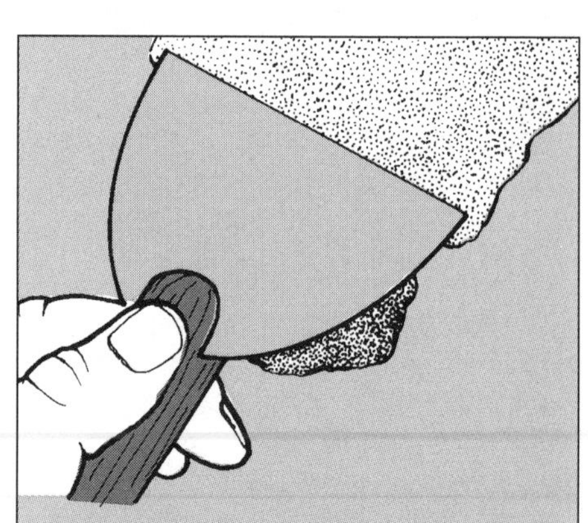

4 The last step is a finishing coat of joint compound. Allow this to dry 24 hours before sanding.

Filling a Crack

1. If the crack is only a hairline, use a can opener to enlarge it slightly. If the crack is wider, follow the instructions for undercutting (see page 9). Gouge out a small bit of plaster at either end to seat the patch. Clear out any dust. If the crack is a long one, use this procedure but gouge holes to anchor the plaster every foot or so along the length of the crack.

2. Fill the crack with patching plaster, overlapping the sound wall. Let dry for 24 hours and fill again if the patch subsides. Let dry another 24 hours and sand smooth.

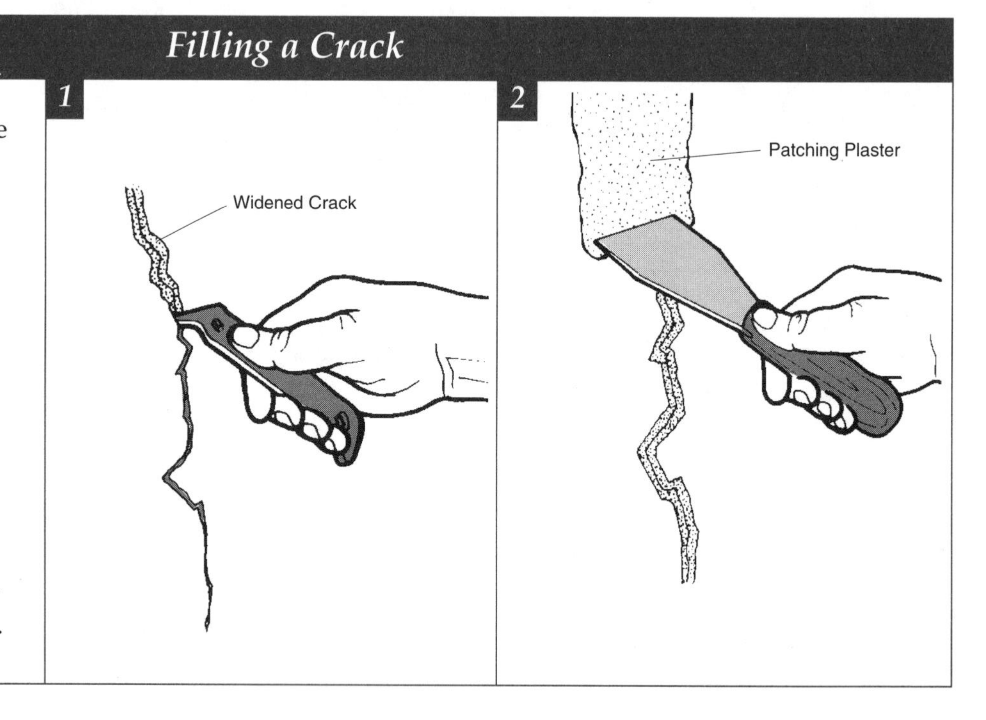

Repairing a Hole in Plaster

1 Clearing Damaged Plaster. Chip and scrape away all loose plaster around the edge of the hole. Any loose plaster will prevent the patch from anchoring to the wall. Do not worry about enlarging the hole.

2 Undercutting the Edge. Give the patch a sound foundation by undercutting the plaster around the edge of the hole. Do this with a can opener or any implement with a hooked point. Carve under the edge so that the plaster slopes away from the edge toward the lath.

3 Filling the Hole. Moisten the edge of the hole with water. Fill the hole with patching plaster, covering the entire surface of the lath evenly, out to a level just below the level of the surrounding wall. Score the plaster with the corner of a spreading knife to make it easier for the top coat to adhere. Let the patch dry.

4 Completing the Patch. When the plaster is dry, apply a coat of joint compound over the patch and feather the edges into the surrounding wall. Allow this coat to dry overnight and sand smooth or apply another coat if the first has subsided below the surface of the wall.

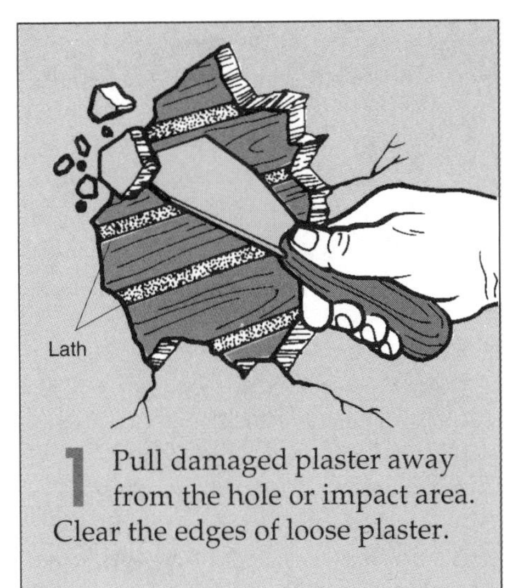

1 Pull damaged plaster away from the hole or impact area. Clear the edges of loose plaster.

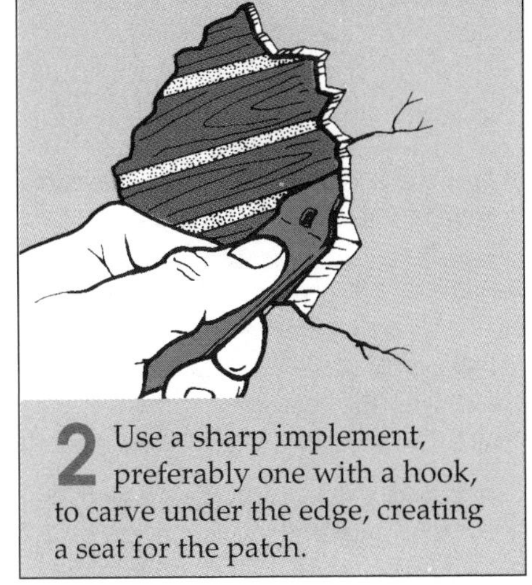

2 Use a sharp implement, preferably one with a hook, to carve under the edge, creating a seat for the patch.

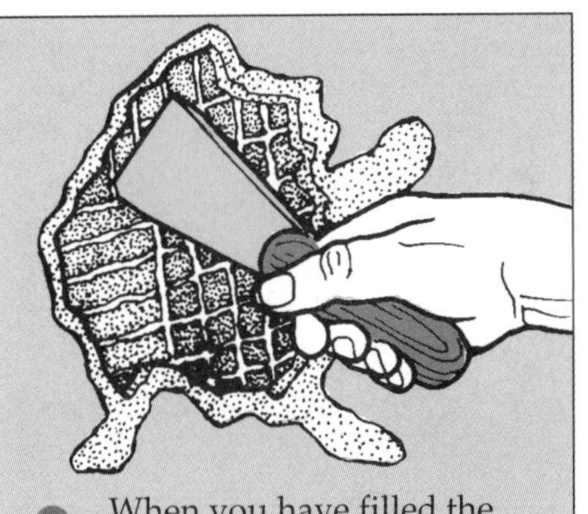

3 When you have filled the hole with patching plaster, groove the surface with the edge of the spreading knife.

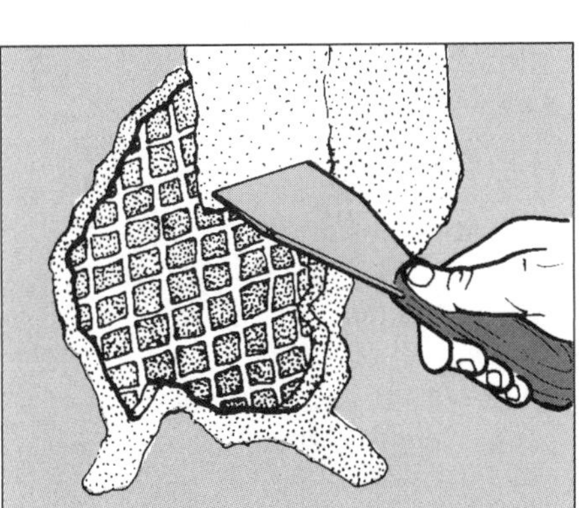

4 Spread joint compound over the patch, filling it flush with the surrounding wall, then feather the edges.

Repairing Plaster at an Outside Corner

If the damage is minor, repair it with joint compound and spreading knife. Where the damage is extensive, tack a straight-edged piece of wood to one side of the corner and use as a guide. Tack far enough above and below the damaged area to avoid causing further damage and gently enough not to fracture good plaster. Fill one side of the damaged area with an undercoat of patching plaster. Use a wooden float to smooth the patch away from the guide toward the good wall; overlap 1/8 inch of the surrounding surface. Move the guide and repeat on the other side. Repair nail holes with joint compound.

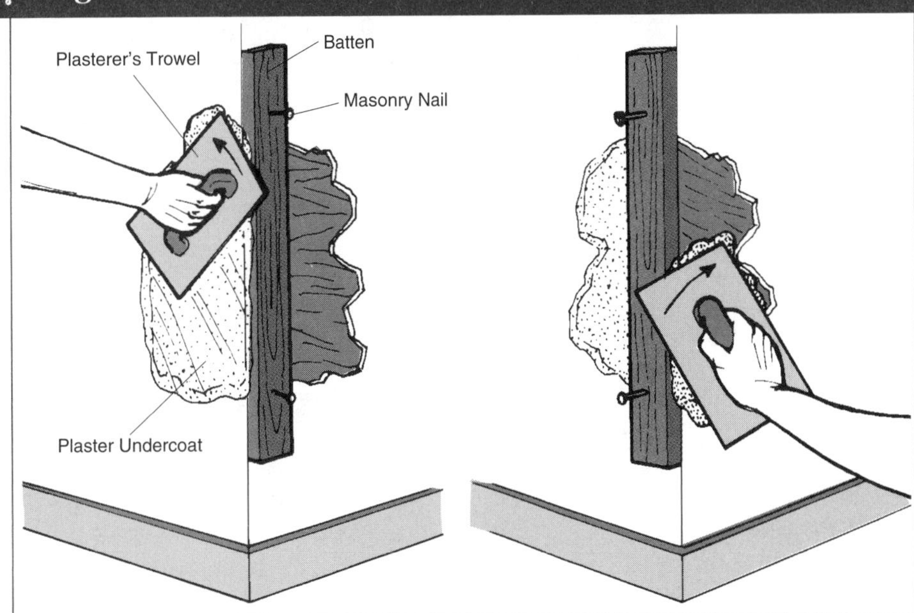

Preparing Walls for New Surfaces

Before installing a new surface, such as paneling, tile or new wallboard on an old wall, you will have to prepare the wall to receive it. This task can range from simply removing molding and cleaning the wall, as in the case of frame wallboard walls that are true and without serious surface defects, to building an entire new wall over an old one that cannot be repaired.

To receive a new surface, a wall must provide a plane of sound nailing or gluing surfaces perpendicular to the floor. A plumb wallboard wall in good condition is the ideal surface to panel or tile—equivalent to a newly built wall. Panels are nailed to studs through the wallboard, or glued with panel adhesive and nailed; tile is glued to the wallboard. Walls with surface defects such as crumbling plaster or unevenness, (i.e. concrete or cinder block walls) must be furred out with a lattice of wood to take nails or adhesive. Framing a false wall is called for when you need space for insulation, wiring or pipes. It is useful in basements and garages. A badly deteriorated surface may have to be removed (see page 14).

The materials you need depends on the wall surface. For furring, use 1x2s and 1x3s which are attached to concrete or cinder blocks with screws into plugs, masonry nails or adhesive, and 8d nails for framing walls. If you must level furring with shims, use wood shingles. False walls are constructed of 2x3s or 2x4s depending on the height of the wall and the weight of the new surface to be attached. Basement walls may require some kind of waterproofing treatment. Basic carpentry tools will take care of all the procedures except driving nails or holes into concrete or cinder block for which you need either a baby sledge and a star drill or a masonry bit on a heavy-duty power drill.

Surveying the Job

On a frame wall, use a carpenter's level to check for plumb at several places along the wall. Then check it for bulges and depressions by running the edge of a long, straight piece of lumber over the wall. A light placed along the edge will reveal low areas; a high spot will lift the edge and cause the board to rock. Isolated low spots can be rebuilt with joint compound or plaster. Isolated high spots on a plaster wall can be knocked down with a hammer. Widespread unevenness must be furred. If you fur a wall, you also must reposition the molding around any windows and doorways (see page 12). Decide whether you will install new molding around the floor and ceiling or reuse what is there. If you are covering a basement wall, decide whether to fur it or frame over it, depending on whether you need space behind the surface.

Moisture Protection

It is important that new surfaces on basement walls be as well protected from moisture as possible. If they are wet, you may need foundation repairs; this is not the time to cover them with a new surface such as paneling or wallboard. Damp walls without significant amounts of moisture should be sealed with a waterproofing paint. To protect panels, cover furring with overlapped sheets of 4-mil polyethylene plastic as a moisture barrier. For further protection from moisture damage, seal the back side of paneling. Leave breathing spaces between furring strips on a potentially damp wall. When you install panels, leave a 1/4-inch breathing space at the bottom.

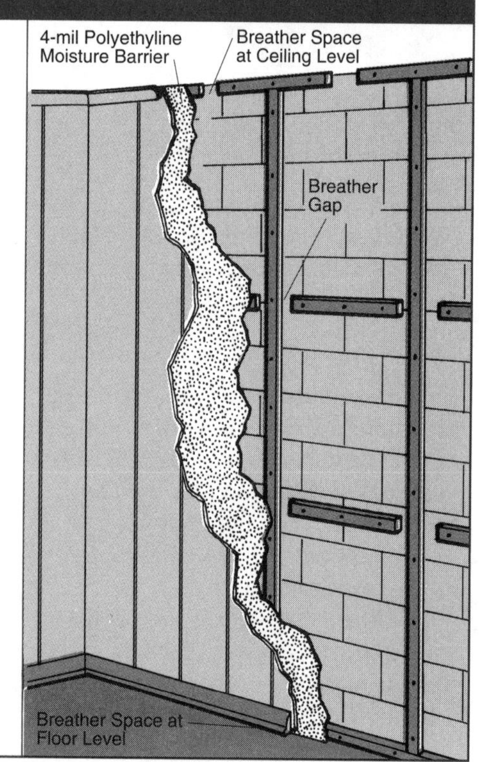

Furring a Wall

Use 8d nails to attach pieces of 1x2 or $1\frac{1}{2}$-inch strips of 1/2-inch plywood to studs. Over a cinder block or concrete wall use masonry nails or screws into plugs. The grid is usually laid out with vertical pieces 48 inches on center and horizontal pieces 16 inches on center. Leave 1/2-inch spaces between pieces where they meet, to allow air to circulate behind the panels.

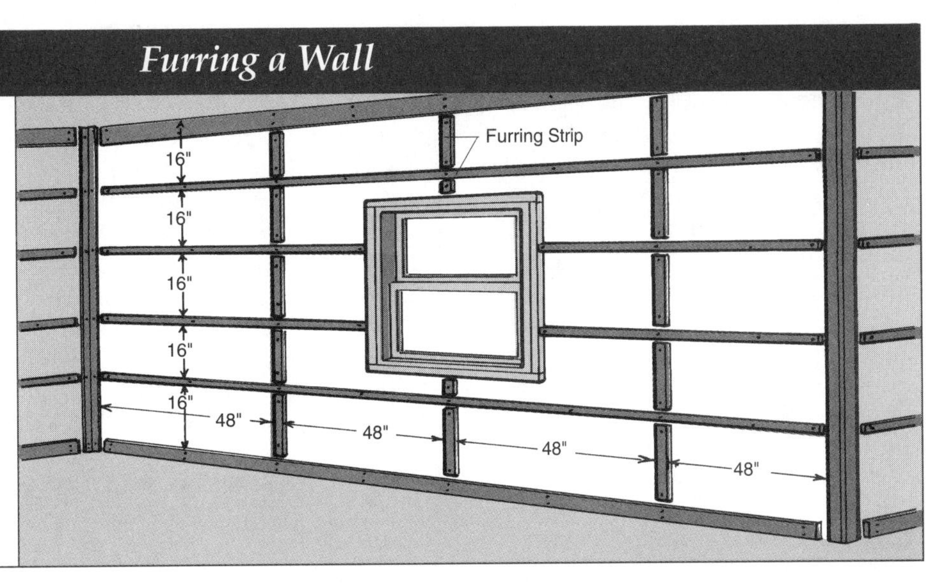

Truing a Wall

1 Measuring the Distance. Use a chalkline to make a grid where the furring will be nailed to the wall and along the ceiling 2 inches out from the wall at each corner. Hang a plumb bob on the ceiling line at the first stud. Measure the distance from plumb line to each intersection of lines and repeat across the grid to find the highest point on wall.

2 Attaching First Strip. Attach a horizontal furring strip at the highest point on the grid. When cutting the strip, allow for vertical 1x3s at the corners. Use single pieces, or butt them over studs. Hang the plumb bob above this point and measure from the line to the furring strip. Then hang the plumb bob from the ceiling line at the first stud and shim behind the furring strip until it is the same distance from the plumb line at the highest point. Nail through the shims into the stud. Repeat at each intersection.

3 Attaching Top Strip. Attach a 1x3 furring strip at the top of the wall by shimming at all points to bring it to the same distance from the plumb bob as the strip below.

4 Completing the Furring. Lay a straightedge over the furring strips to find their correct height. Make the furring flush with the straightedge. Attach vertical pieces in the corner.

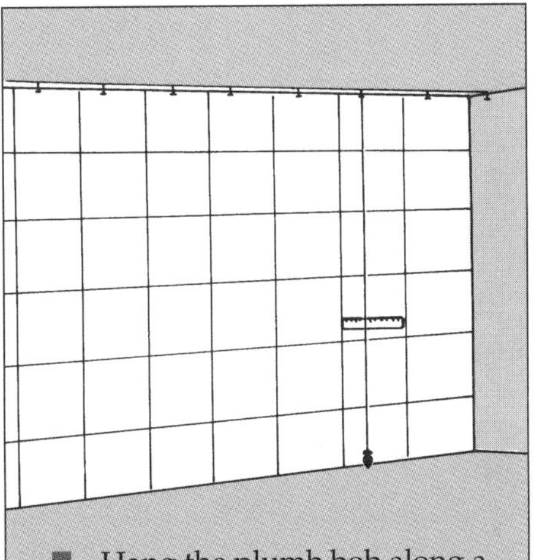

1 Hang the plumb bob along a line on the ceiling 2 in. from the wall. Mark distance from wall at every intersection.

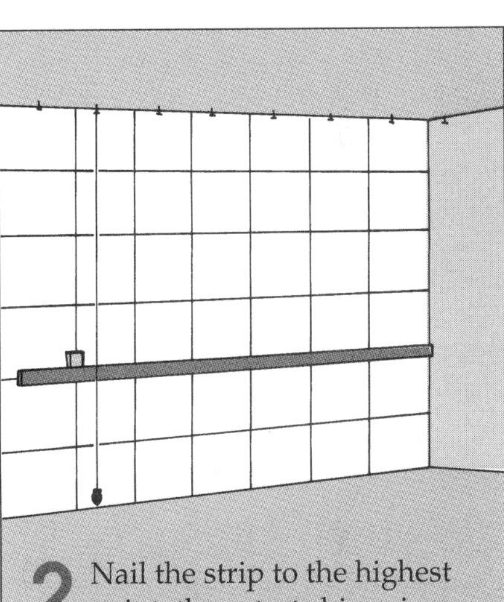

2 Nail the strip to the highest point, then start shimming behind to make it level along its entire length.

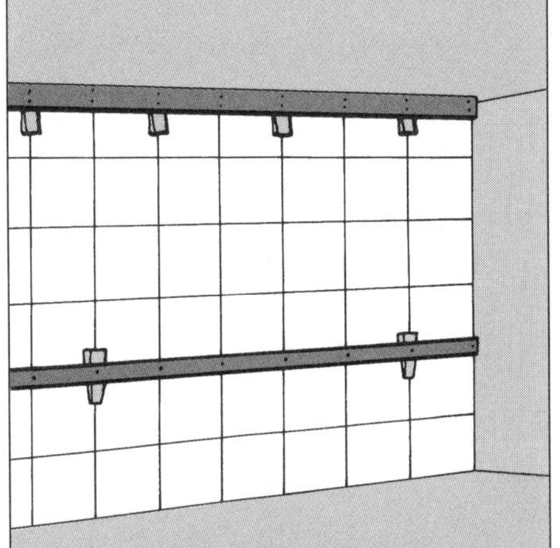

3 When the first strip is in place, attach the top strip the same distance from the plumb bob as the first strip.

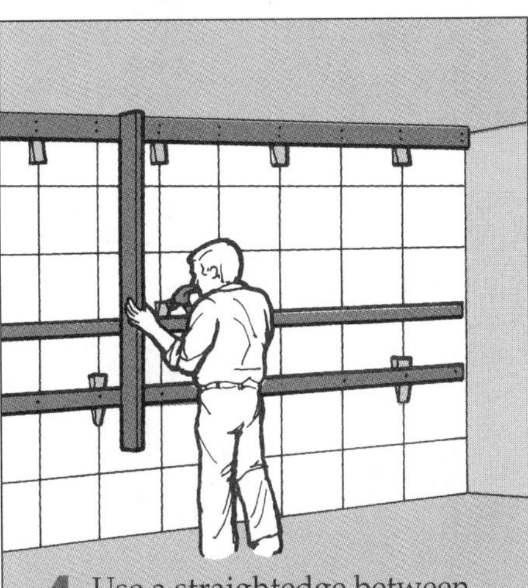

4 Use a straightedge between the top and first strip as a guide for shimming the rest of the strips.

Furring Around Windows

1 Removing Molding. If you want to save the casing around windows, find the nails at the corners and drive their heads through into the adjoining piece with a nail set, then pry the casing away from the wall. If the window has a stool, drive the heads of nails through the horns and pry it out.

2 Adding Furring. Add furring strips around the window on all sides. Because the new surface will deepen the window opening, you will have to add jamb extenders so that molding can be reinstalled on the new surface. The extenders should be the depth of the new surface plus the depth of the furring. Cut a new window stool to fit the depth of the new surface. You can find the dimensions of the new window stool by tracing the old one on new stock. Remember to add the depth of the new wall surface.

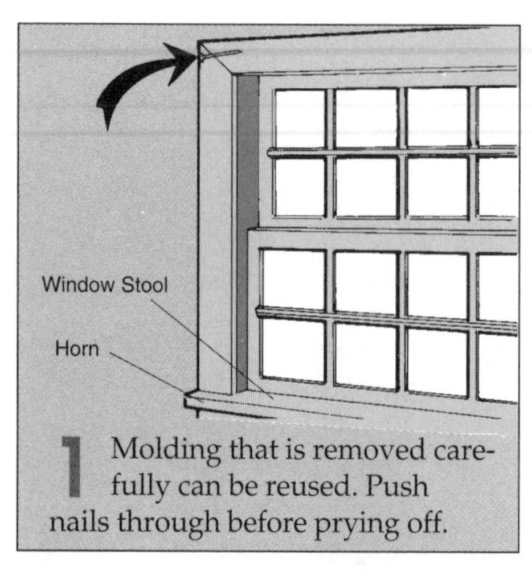

1 Molding that is removed carefully can be reused. Push nails through before prying off.

2 Every place where the new surface will have an edge must be furred.

Framing a False Wall

To make room for insulation or wiring under a new surface fastened to a concrete or cinder block wall, erect a frame of 2x3 or 2x4 studs against the masonry. Plan how you will arrange plates at the corners and cut top and bottom plates the length of the wall (or walls). Mark them for stud locations, 16 inches on center as shown on page 20. Build a frame on the floor as described on page 20. Then raise it into position. Check for plumb and shim at top and bottom where necessary, then attach at top, bottom, and sides—use 16d nails for wood, masonry nails for concrete. Frame around any electrical outlets.

Boxing in Pipes & Ducts

Pipes and ducts that intrude into a room can be framed to receive paneling. To conceal a pipe in a corner, build the two short walls and attach the frame to the floor, walls and ceiling (left). Horizontal pipes at the ceiling are concealed by walls on their sides. Build the two walls and attach them to each other in place (right).

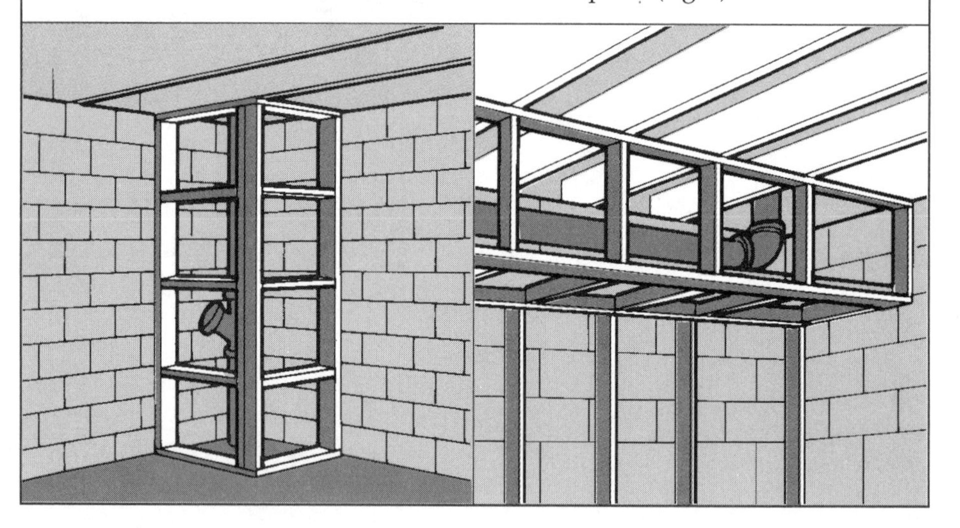

Removing Walls

Removing a wall changes interior space dramatically. With the removal of a partition, two small rooms side by side become a comfortable master bedroom; and a cramped living room adjoining a small dining room becomes more contemporary open space. If you are considering such a change, make a scaled drawing of the new space to be sure it will have agreeable proportions.

There are two kinds of interior walls in every house and their difference is extremely important. Bearing walls are structural supports that carry some of the weight of the story or roof above (see below). Non-bearing walls are room dividers with no structural role at all. The support provided by a bearing wall must be replaced with a beam (see pages 15-16), but a non-bearing wall can be removed easily without structural considerations. The gaps left in abutting walls and ceilings can be repaired by following the directions on pages 7 and 55.

The walls in almost all houses built in the last 40 years are stud frames covered with sheets of wallboard. Older houses (and a few newer ones) have stud walls covered with plaster on a wood or metal lath. You can usually tell the difference by knocking on the wall; wallboard has a hollow sound between studs, while plaster is more solid. As noted in the directions that follow, removing a plaster surface is slightly different from removing wallboard.

Tearing out a wall obviously is messy, so protect furniture by moving it and cover the floor with drop cloths. Because of the dust this job will create, the space should be ventilated. Always wear a filter mask and goggles when tearing out a wall.

Determining the Type of Wall

The most important clue that a wall bears weight is that it runs perpendicular to the joists above. If it runs parallel to the joists, you can be certain that it is a non-bearing wall; if it does not, it may or may not bear weight. If joists are overlapped above a wall (as shown below) then it is a bearing wall. Joists which are not overlapped are probably supported by a wall that is near the middle of the distance they span. For instance, if a wall, such as a closet wall, is near one end of the span it is probably non-bearing, but consider what is above. For example, the wall may be bearing the weight of a concrete bathroom floor. Studs set closer together than the standard 16 inches may indicate that a wall bears weight.

A beam or a wall in the basement running under the length of the wall in question is another indication that it bears weight. If none of these points of inspection is available, cut a peephole in the ceiling (see below) to see what is above. If you have any doubts, assume that the wall bears weight, or consult an architect.

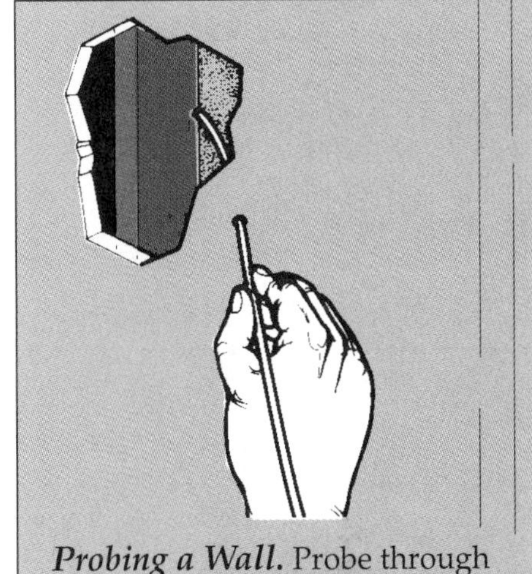

Probing a Wall. Probe through a drill hole with a stiff wire. When you hit an obstruction, hold the wire at the wall, pull it out, and measure that length.

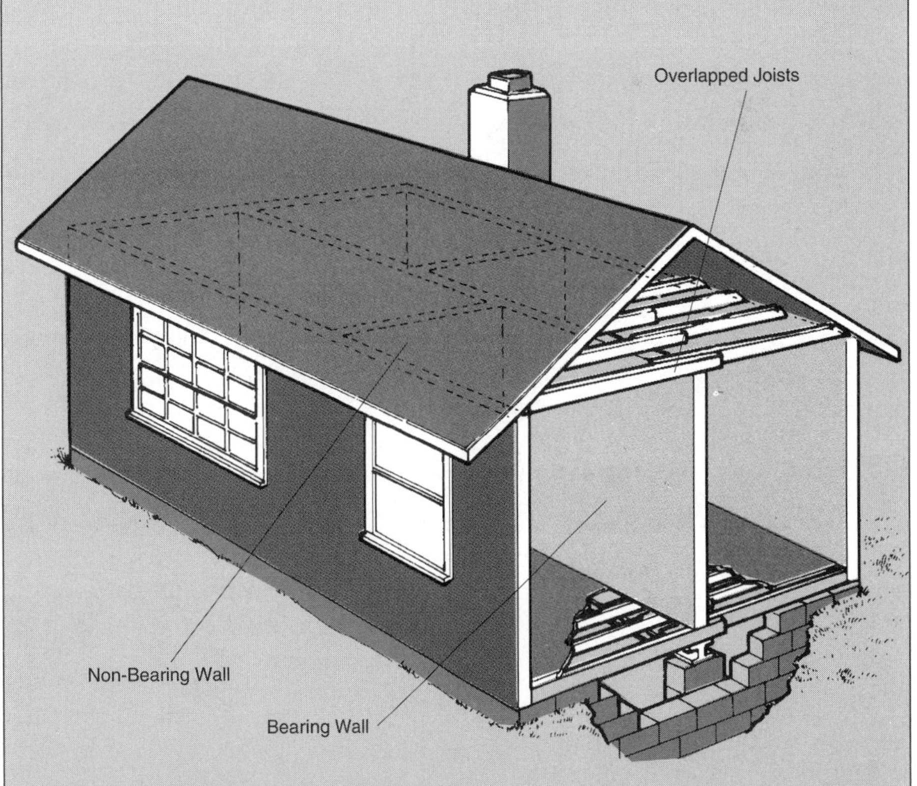

Determining the Type of Walls. An unfinished attic or crawl space above a wall makes it easy to determine whether the wall is bearing or nonbearing. Look at the direction of the joists above the wall. Joists perpendicular to a wall usually indicate that it bears weight; overlapped joists always do.

Removing Non-Bearing Walls

1 Rerouting Wire. Trace pipes and ducts by checking where they pass through the floor in the basement or emerge in a second story or attic. When in doubt probe through holes in the wall (see page 13). Shut off power to the entire area of the house where you are working and use a heavy-gauge extension cord to bring power from another part of the house.

2 Removing Trim. Pry off any trim. If you take care not to damage it, you can reuse it. Remove all outlet and switch face plates. Tape drop cloths to floor to catch debris.

3 Cutting the Wall. Before beginning work, heed the caution found on page 6. If the wall is free of obstructions like wiring, use a circular saw to cut out the wallboard between studs. Cut the studs in half and wrench out the pieces. For plaster over metal lath, use a metal-cutting blade. Plaster also can be removed by scoring the surface with a saw blade set just to the depth of the plaster, and then knocking out the chunks.

4 Finishing the Job. The last stud is usually nailed to two studs set together or to blocking between two close-set studs. Pry it from the bottom. Pry the top plate from its nailers starting at one end and using scrap to protect the ceiling. Cut a few inches out of the middle of the sole plate and pry up the two halves from the cut.

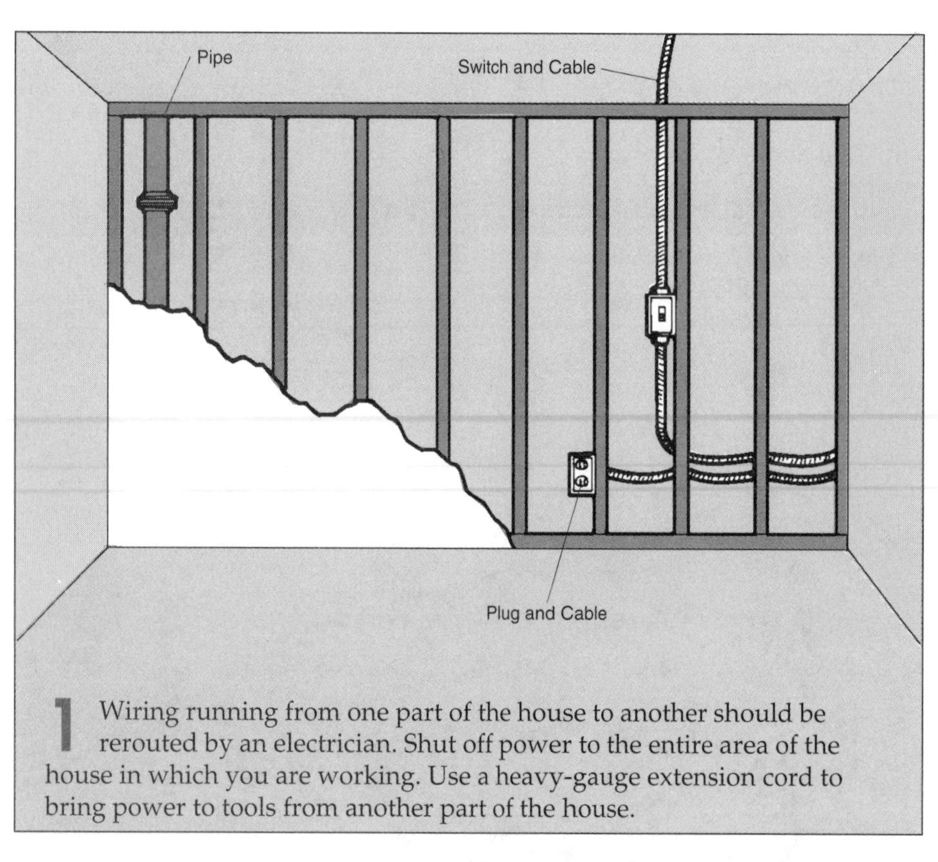

1 Wiring running from one part of the house to another should be rerouted by an electrician. Shut off power to the entire area of the house in which you are working. Use a heavy-gauge extension cord to bring power to tools from another part of the house.

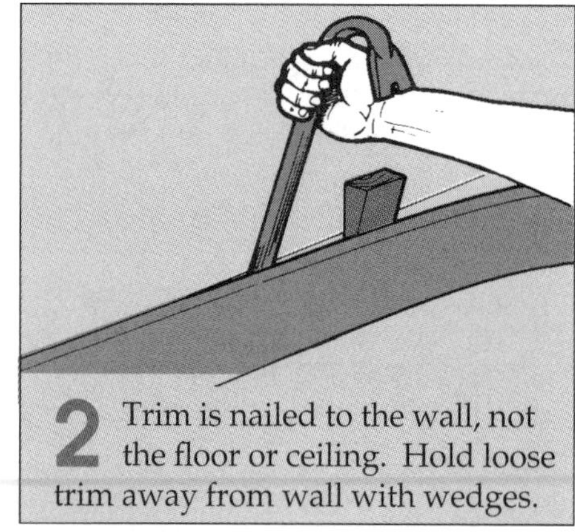

2 Trim is nailed to the wall, not the floor or ceiling. Hold loose trim away from wall with wedges.

3 Avoid hammering surface of studs, which damage ceiling and floor.

4 To pry the last stud, use a piece of scrap to protect the good wall from the crowbar.

Shortening a Wall

Remove the wall to the stud where you wish it to end, but cut the top plate and sole plate $1^1/_2$ inches from the last stud. Cut into the ceiling so that you have working access to both sides of a pair of joists, add a 2x6 nailer between the joists and nail the top plate to it.

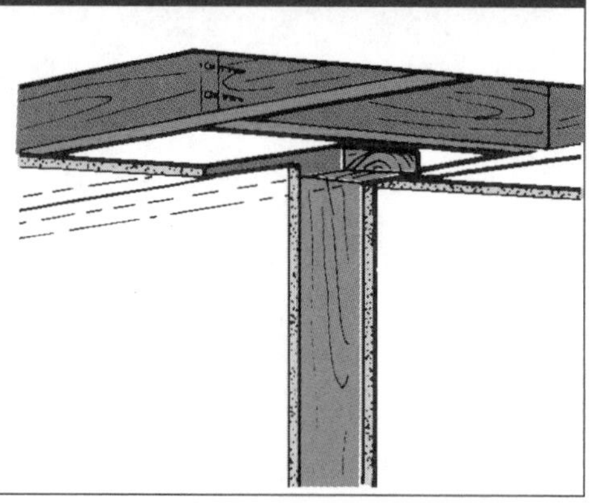

Removing Bearing Walls

If you have determined that a wall you wish to remove is a bearing wall (see page 13), you must replace its load-carrying function with a beam. Either a wooden header or a steel girder may be used, depending on the width to be spanned.

The studs in a bearing wall carry the weight above the wall to the joists below. Because the studs are set relatively close together, the top plate need not be particularly strong. But when the wall is removed, the beam that replaces it must be strong enough to bear the weight along its full length to the posts at either end. For spans up to 8 feet, wooden beams, either solid or laminated, may be used. For spans beyond 8 feet, steel girders must be used. No matter what distance is spanned, these beams will protrude below the ceiling, but they can be covered with wallboard and finished to match the ceiling. False beams added next to working beams conceal their presence (see page 73).

The weight carried by a bearing wall is distributed among the several studs so that no one stud concentrates great weight on the joists beneath. But when the total weight is borne by only two posts at either end of a beam, it is quite concentrated, and you must be especially cautious that there is sufficient support beneath to carry it. For spans greater than 8 feet, it is wise to consult an engineer to determine what extra support you may need. Spans greater than 14 feet concentrate so much weight on the posts that they should be handled by a professional builder.

An entire bearing wall between two adjoining walls can be replaced if the distance is manageable, or replace part of a bearing wall by butting one end of the beam to the end of the foreshortened bearing wall. Another solution is to create an arch by removing part of the wall and butting the beam to the remaining wall on either side.

1 Building Temporary Supports. Order or cut a beam 7 inches longer than the span you will be bridging. For ease of handling later, place the beam beside the length of wall it will replace. Build two stud walls the length of the beam, one on either side of the wall, see pages 19-21. Nail a 1x4 brace diagonally across each stud wall from corner to corner, attaching it to every stud. Install these temporary supports about 2 1/2 feet from either side of the wall by driving wedges between the top plates and each joist above. Check as you work to be sure the wall is absolutely plumb. You do not have to nail these temporary supports in place; the wedges must hold the stud walls tight. Do not proceed unless they do.

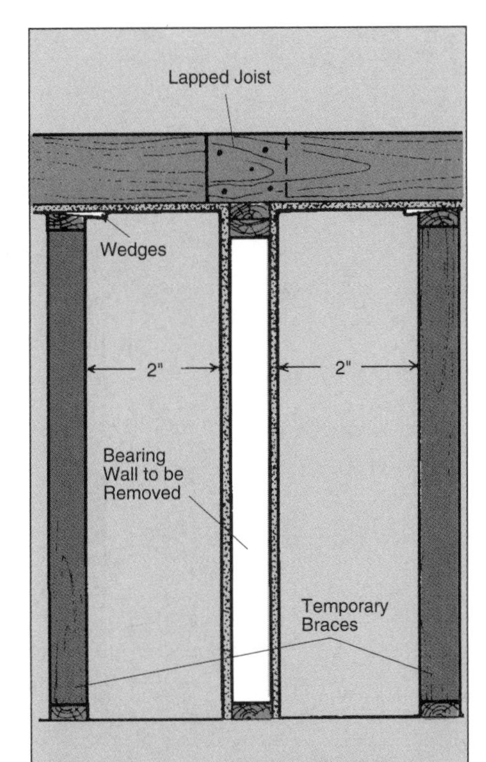

1 Stud frames on either side of the bearing wall you are removing support the weight above until a beam is installed. They must be perpendicular to the floor and ceiling, and shimmed tightly at the top.

Types of Beams

Three types of beams are common in home construction jobs of this kind: solid wood, laminated wood and steel I-beam. Steel is used for spans greater than 8 feet. The size of the wooden beam you need depends on the distance you will bridge, as shown in the table below. Laminated beams are far less expensive than solid wood, and are the choice of most home-owners. They are made by putting a piece of 1/2-inch plywood between two boards and fixing this sandwich with 16d nails driven through one side. Solid wood beams should have the same dimensions as laminated beams.

Span	Lumber
3 1/2 ft.	2x6
3 1/2- 5 ft.	2x8
5 - 6 1/2 ft.	2x10
6 1/2- 8 ft.	2x12

Steel I-beams for home use are called 8x17 steel girders, meaning they are 8 inches deep and weigh 17 pounds per foot. They can be purchased from steel suppliers.

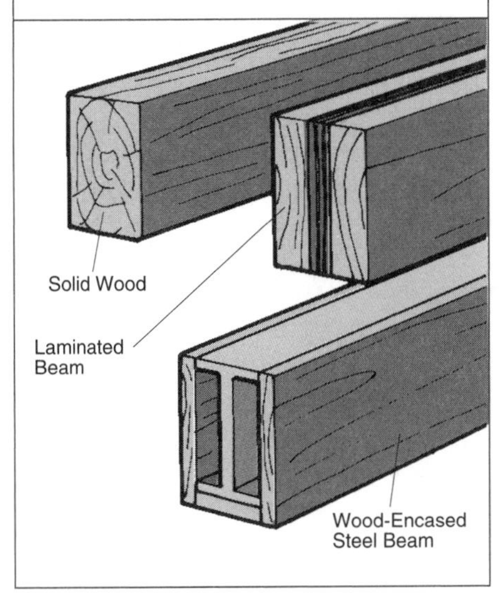

2 Dismantling the Wall. When the temporary walls are securely in place, dismantle the bearing wall following the directions for removing a non-bearing wall, page 14. Because the top plate in a bearing wall is a doubled 2x4 tied into the walls it meets, cut a few inches from the center with a saber saw, pry down the two halves from the cut, and wrench them free. If you are installing a steel I-beam, do not remove the top plate. Remove the surface of the wall (or walls), abutting the wall you are dismantling as far as the studs on either side of the intersection.

3 Adding Extra Support Beneath the Floor. If you are working above a basement with an unfinished ceiling that gives access to the floor joists, check to be sure the sole plate of the adjoining wall, on which the post will stand, is sitting on a joist above a beam. Do this by driving a nail through the floor next to the plate and locating where it has come through below. When the adjoining wall is between joists, add extra support. If you cannot determine this, or if there is no beam below, consult a professional builder.

4 Installing the Beam. If you are installing an I-beam, see page 17. Notch a wooden beam 3 inches deep and 3½ inches in from the end at either end that will meet an adjoining wall. Leave it plain where it will abut a remaining segment of the bearing wall. Place the beam on the sole plate of the adjoining wall, notch side up, and measure from the notch to the top plate above. Cut a 4x4 post exactly that length and check for fit. With help, raise the beam into position. If sagging joists prevent it from meeting all the joists flush, raise the joists by driving extra wedges between them and the temporary walls. Seat the posts under either end of the beam and toenail to the sole plate and beam. Add 1x4 nailers on either side.

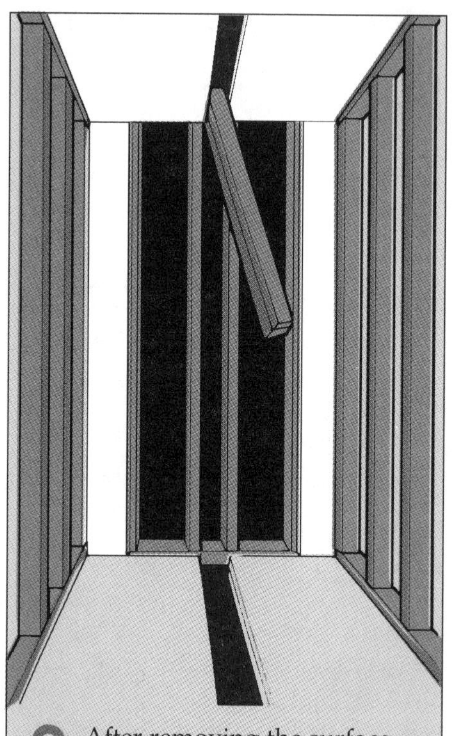

2 After removing the surface, studs and sole plate, cut the top plate in half and remove the pieces.

3 Posts supporting beams may need the extra support of 2x6s nailed between the floor joists.

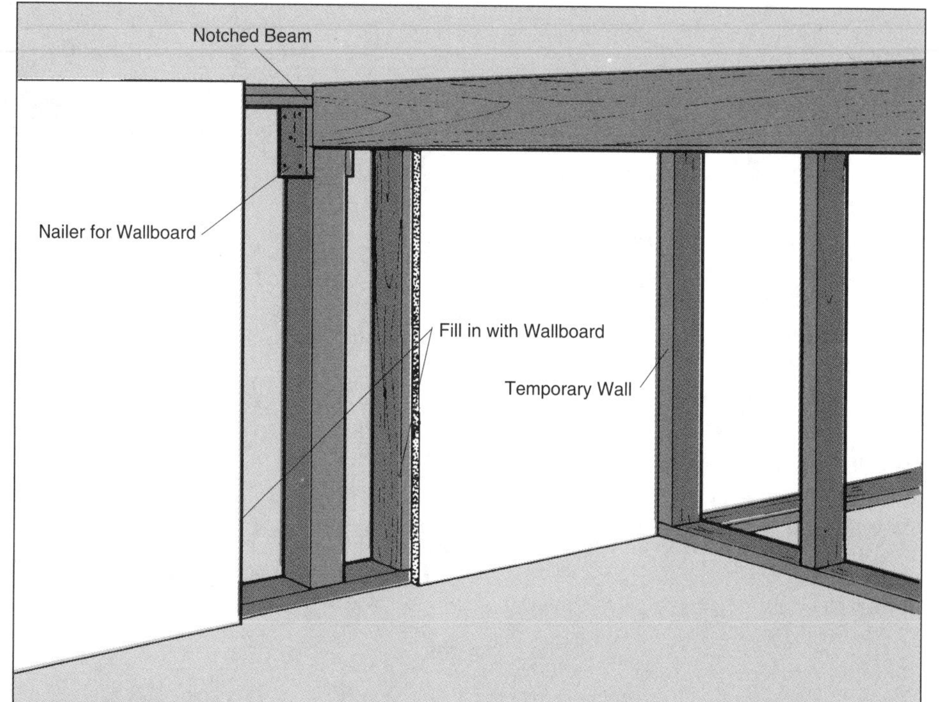

4 With the beam in place and the posts secured, add pieces of 1x4 flush with the forward edges of the posts to provide a nailing surface for wallboard. The gap left in the adjoining wall(s) is filled with a new piece of wallboard cut to fit between old studs.

Butting a Beam to a Wall

To butt a beam to the remaining section of a bearing wall, use the procedures described on the preceding page, but leave $3^1/_2$ inches of the sole plate extending from the last stud of the remaining wall. Cut a post the length of the distance between the beam resting on the sole plate and the bottom of the joist above. Put the beam in place against the last stud and drill through both the post and the stud at four points—near the top, bottom and two in the middle—for long toggle bolts. Use a bit the same size as the bolt. Install the post (or posts) with the bolts as shown.

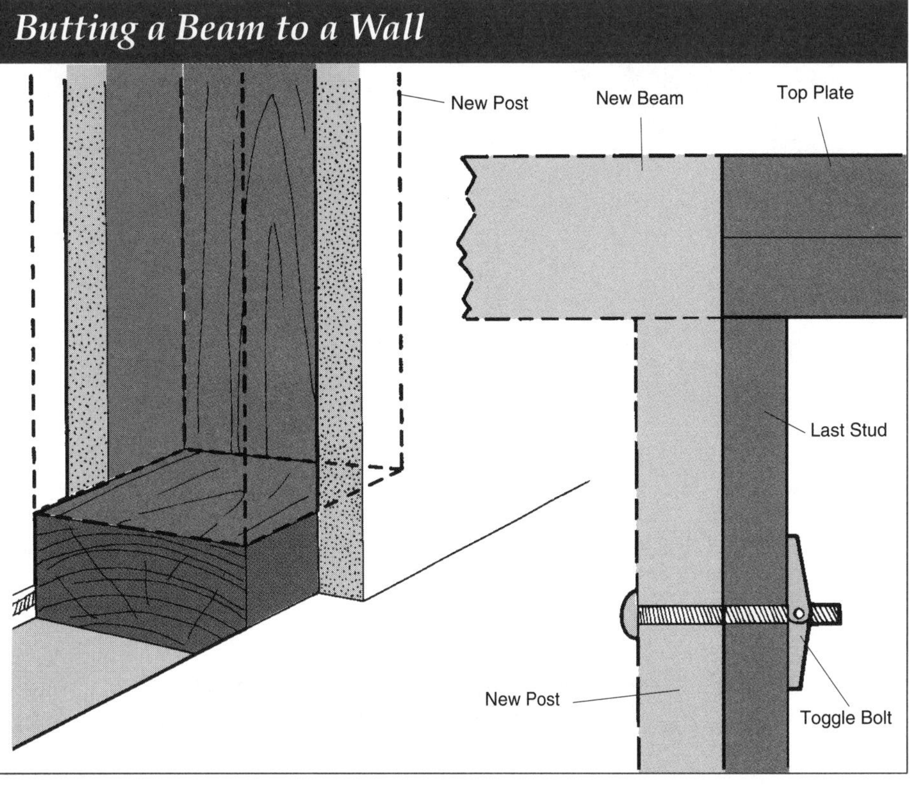

Installing a Steel I-Beam

1 Raising the Beam. Installing a steel beam is not much different from putting up a wooden one, but it is heavier and requires more help. Plan to have one able-bodied assistant for each 3 feet of beam when you raise it. Another difference is that the metal beam sits against the top plate of the wall it replaces, so do not remove the top plate when taking out the bearing wall.

The I-beam should be 6 inches longer than the distance it will span. When the wall is removed (with the top plate still in place), lift the beam to shoulder height and position stout stepladders under both ends to rest the beam on. Raise it to the top plate and check that it fits flush from end to end. If the top plate is bowed, set the beam on the ladders and tighten the shims in the temporary walls under the low points to level the plate. Fit the beam against the leveled plate and install posts under either end. If you are butting the beam to a wall, follow the instructions in the box above.

2 Finishing the Beam. Cover the beam with nailing surfaces for wallboard or paneling. Use 1x6 lengths on the bottom of the beam, 1x2 lengths for the vertical strips, and boards cut to fit to fill between the beam and the ceiling. Cover the beam with wallboard or paneling.

1 The posts that support an I-beam should fit tightly When the beam is flush against the ceiling; set the posts.

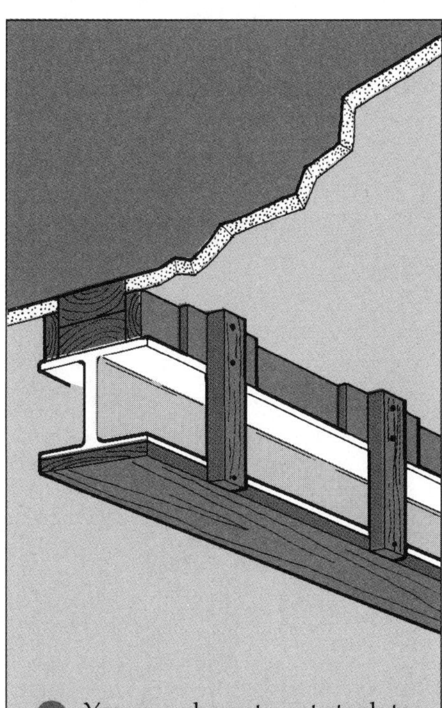

2 You may have to cut stock to fit between the beam and the ceiling to provide support for the vertical nailers.

Framing Walls

Putting up a wall involves three separate activities: building the frame, covering the frame, and finishing the surface. The frame is a rigid skeleton of studs that provides space for insulation and a foundation for wall surfaces. With attention to a few simple procedures, anyone who can take accurate measurements, hammer a nail and saw a straight line, can frame walls that are sturdy and true.

A frame must be vertical and rigid, and it must fit tightly with the floor, ceiling, and walls it meets. All these conditions can be assured by accurate measuring, marking and cutting. Unfinished basements are usually the easiest places to frame walls because the walls, floors and ceilings already in place probably meet at right angles and are exposed, ready for the attachment of new framing. Because houses settle, rooms on the ground floor and above seldom offer right angles all the way around and this complicates putting up new walls. Also, the job often involves breaking into finished walls or ceilings or both. (An unfinished attic presents special problems; see page 22.) Where the space you are framing is true, you can build most of the wall on the floor and erect all the studs in one motion. Where the angles are off, you must put up the wall stud by stud. The two methods are treated separately in the directions that follow.

Wood, usually 2x3s or 2x4s, is the most popular framing material, but there is an aluminum alternative that offers most of the qualities of wood and has such advantages as nonflammability and greater ease of handling (see page 24).

The first stage in building a frame wall is to determine how it will be attached to the ceiling and adjoining walls (see below and next page). Find the location of joists above the wall by the methods described on page 13.

Attaching Walls to Ceilings

Across Joists. If the joists above run perpendicular to the new wall, the wall is simply nailed across the joists at each one. If the ceiling is finished, fix the top plate to the joists by nailing through the ceiling. The seam between the surface of the new wall and the ceiling can be concealed by molding (see page 38).

Along Joist. If your new wall falls along a joist, nail the top plate through the ceiling to the bottom of the joist along its length as described above. Check the joist at both ends of the new wall and see if it runs true along the line you mark for the wall. If it does not, try to position the wall so that you have a sound nailing surface against the joist at both ends. Mark the top plate so that the nails will go into the center of the joist.

Between Joists. If you place a new wall so that it runs parallel to, but between joists, you must add nailing blocks between the joists so the top plate can be attached to something. To do this, strip the ceiling on either side of the two joists between which you will add nailing blocks. Cut 2x4s to fit between the joists and end nail them flush with the bottom of the joists. Repair the ceiling with wallboard cut to fit the gaps along either side of the new wall.

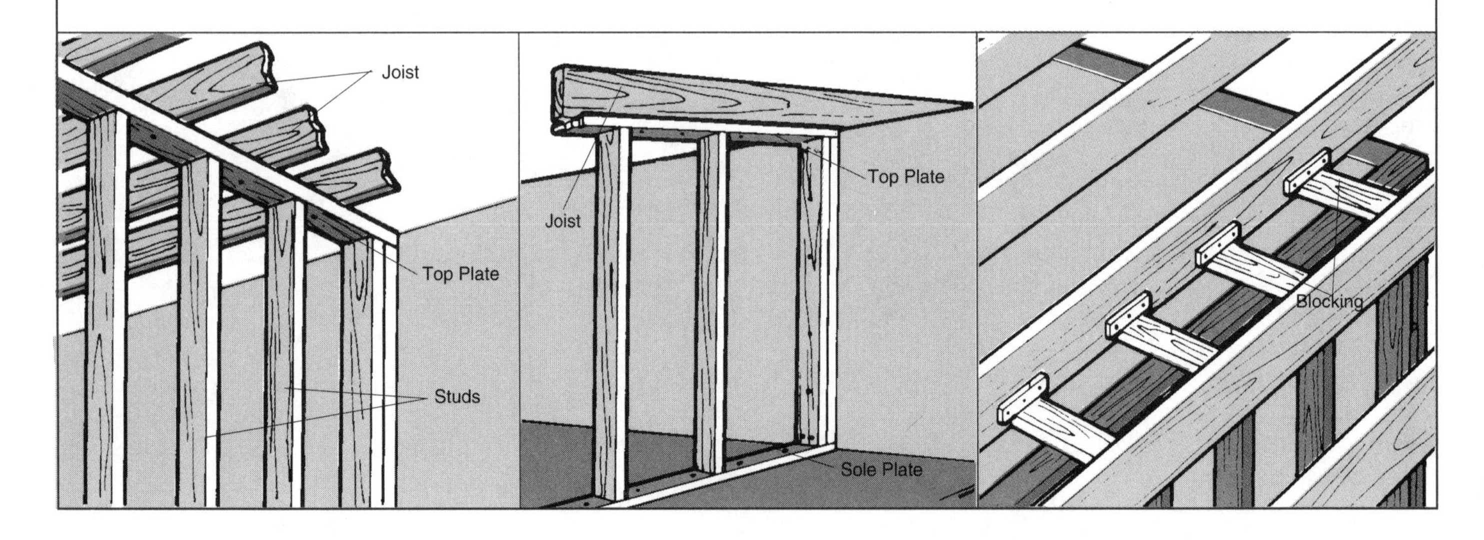

Attaching Wall to Adjoining Wall

At a Stud. Locate studs in adjoining walls by the methods shown on page 13. If a new wall meets an adjoining wall at a stud, remove sections of trim above and below and attach the new wall to the stud directly through the old wall's surface.

Between Studs. If the new wall meets an adjoining wall between studs, strip the wall surface back to the studs on either side of where the new wall will meet it and install a double stud with 2x4 blocking in the middle. The double stud is necessary to provide nailing surfaces for the new wall and for repairs to the old surface where you have broken through it.

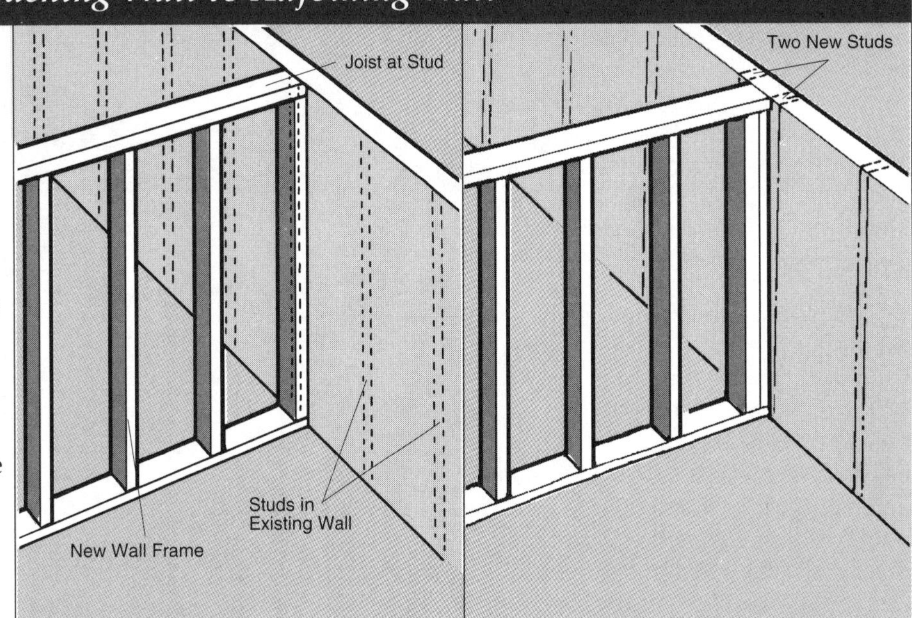

Hanging a Stud Wall

1 Planning Ahead. The height of the wall you plan to build will determine the choice of lumber. You can use 2x3s for walls up to 8 feet high, 2x4s for walls higher than 8 feet, or you may wish to use larger lumber to improve sound-proofing. You will need one or more studs at each end of the wall, depending on how you tie it to old walls, and studs every 16 inches between. When planning your lumber purchase, account for any double studding needed around a door.

2 Marking the Location. Mark the exact location of the wall with a carpenter's square and chalk line. If the wall is square, lay one leg of the square against it and have a helper line up the chalk line on the other, then snap the line.

3 Attaching the Sole Plate. Measure the full length of the new wall across the ceiling. Cut two pieces of lumber (for a wood floor), or three pieces (for the double sole plate needed over a concrete floor). Nail down one length along your mark, if installing a double sole plate. If not, do the next steps first, then attach the sole plate.

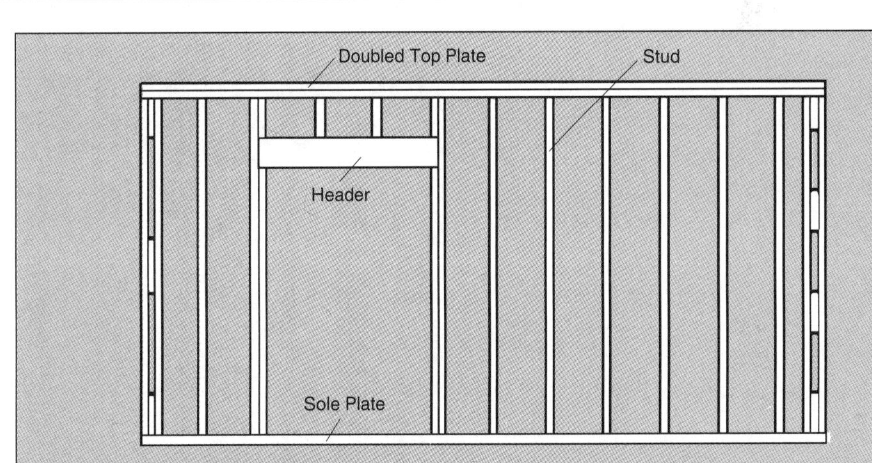

1 For walls up to 12 ft. long, use a single board for each plate. For longer walls use two pieces for each plate with no piece shorter than 4 ft. and with splices at opposite ends of the wall. Walls up to 8 ft. high can be built with 2x3s; for walls higher than 8 ft., use 2x4s.

2 Locate the place where you want the wall, then stretch and snap a chalk line as a guide.

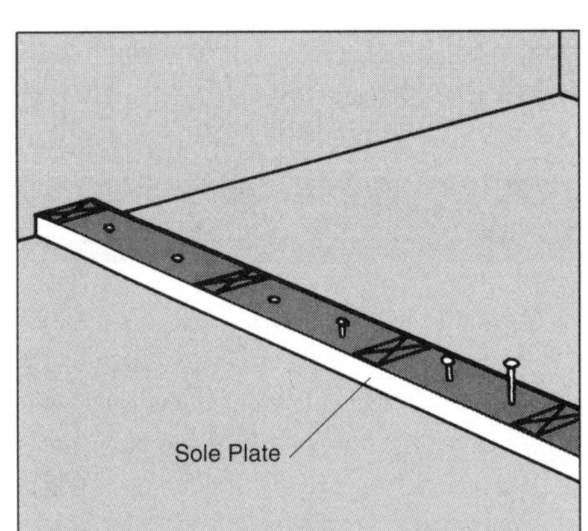

3 Nail the sole plate to a concrete floor with masonry nails, to joists with 16d nails.

4 Marking the Plates. Align the top and bottom plates as shown. Measure from one end of the plates and mark for studs at 16-inch intervals along the plates (some building codes permit 24-inch spacing) and at the other end even if it is less than 16 inches to the last stud in the sequence. Remember to account for the doorways.

5 Building the Frame. Set the plates on edge with the marks facing each other, separated by the height of the wall. Put the studs in position and nail top and bottom with two 12d nails through the plates into the ends of the studs. Use your marks to align the studs precisely so the frame will be true and provide dependable nailing surfaces.

6 Raising the Frame. Raise the frame by setting one plate on the plate already attached to a concrete floor (or on your marks on a wooden floor). Then, walk the frame into position. Position it above on the marks on the joists or nailers. If the fit is exact, you may have to tap the top plate into position. If the fit is tight, attach the bottom plate with 16d nails. If it is loose, go to the next step first.

7 Shimming the Frame. Use wood shingles as wedges to make the frame fit snugly if it is loose anywhere. Pound the shingles beneath the bottom plate before attaching the frame at the bottom. Use a carpenter's level to check that the frame is plumb—check both the outside and inside faces of a few studs. Adjust the frame as necessary and attach the top plate to the joists with 16d nails.

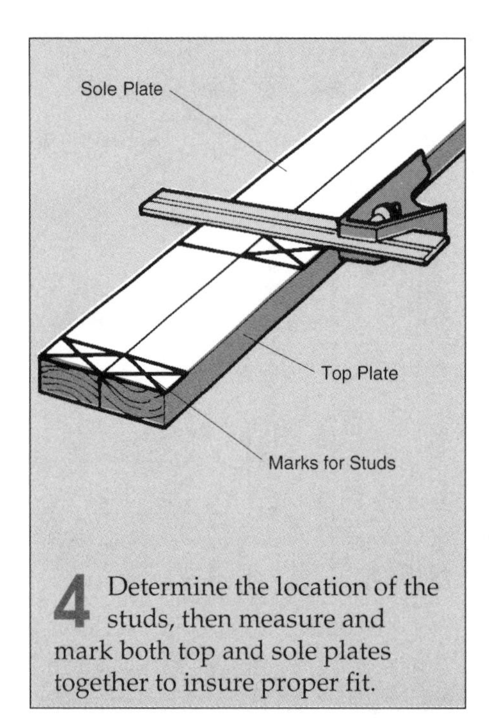

4 Determine the location of the studs, then measure and mark both top and sole plates together to insure proper fit.

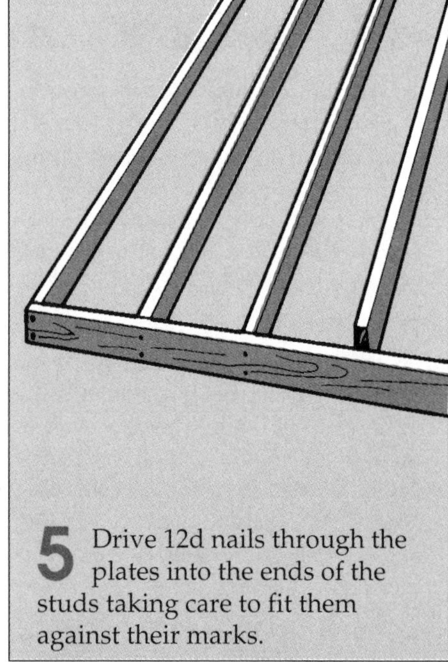

5 Drive 12d nails through the plates into the ends of the studs taking care to fit them against their marks.

6 Get some help to raise the frame if it is large. You may have to hammer it into position.

7 If the frame is loose against the ceiling, drive shims under the bottom plate or over the top plate.

Turning a Corner

At a corner you must add an extra stud to provide a nailing surface for wallcovering. One method of turning a corner involves nailing spacers between two studs, then butting the end stud of the adjacent wall to this triple-width stud. Another method uses three plain studs as shown right.

Plate
Extra Stud
End Stud

Plate
Extra Stud
2x4 Spacer
End Stud

Building Stud Wall in Place

1 Attaching the Top Plate. Where the length of the wall, or a room out of square prevents you from building a stud wall on the floor and then raising it into position, you must build the wall in place. First mark the position of the wall on the ceiling, then cut top and bottom plates to fit. Install the top plate.

2 Locating the Studs. Use a plumb bob to find the location of the bottom plate. With a helper to steady and read the plumb bob, mark the floor for the bottom plate. Move along the top plate, mark the floor at the other end, and check it in the middle. Attach the bottom plate according to the procedures described above. Then mark the top plate for the position of the studs at 16 or 24 inches on center, and outline the stud (3/4 inch from center on either side). Use these marks and the plumb bob to locate the position of the studs on the bottom plate. Be sure that the outlines of the studs are oriented the same way (top right-hand corner to top right-hand corner, for instance). Outline the studs on the bottom plate. Measure from a stud location on the top plate to the corresponding location on the bottom plate and cut a stud that length—the fit should be snug, but not so tight that the stud bows.

3 Toenailing. Put the stud in position on the marks and attach by toenailing at the top and bottom with 16d nails. The easiest way to toenail is to start a nail on either end before you raise the stud; drive the nail at a 45-degree angle into the middle of the face of the stud, until it just breaks the surface on the end. Put the stud in position and complete the attachment by toenailing with two nails into the opposite face at the top and bottom. Use a spacer block, cut to the exact distance between the edges of the studs to make toenailing easier.

1 If the ceiling is unfinished, nail across or along joists, or in toenailers between joists. On a finished ceiling, locate joists through the ceiling. Adding nailers between joists requires opening the ceiling.

2 Mark off stud locations on the top plate and use the plumb bob to find the same positions directly below on the bottom plate. Mark those positions. Studs should be cut individually to measurements between top and bottom plates at stud locations.

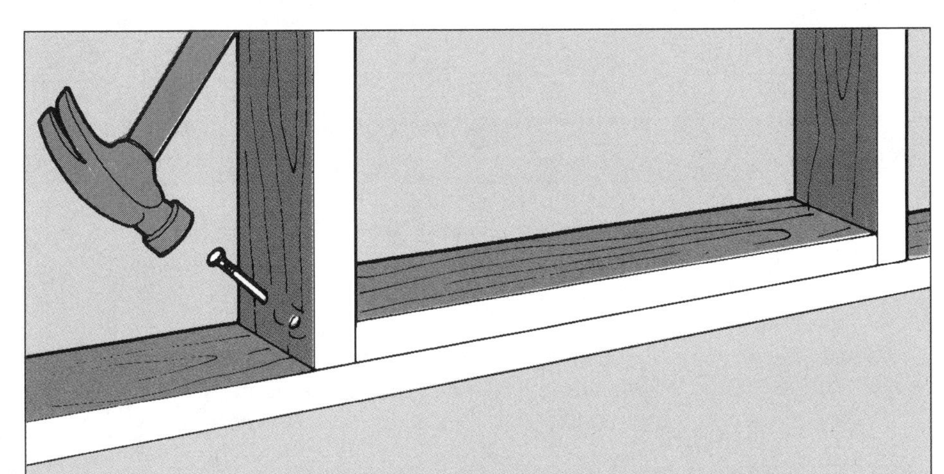

3 Start a nail through either end of the stud then put in position. Use a spacer to hold the stud in position while you nail. If you cannot use a spacer, start nailing with the stud slightly in front of its location so it will slip into the proper position as you nail.

Framing-In an Attic

Framing to finish attic spaces under a sloping roof requires methods somewhat different from framing-in a space with a level ceiling. The two major differences are that some surfaces must be provided between rafters to which studs can be attached for a wall running between the sides of the attic; and the walls along the sides will be low—usually only 4 to 5 feet high—and will meet the rafters or a plate at an angle other than 90 degrees.

A wall running across an attic is fixed at the top to a collar beam installed between rafters at the desired ceiling height. The top plate of a stud wall can be nailed to the collar beam. The shortened walls along the sides are called knee walls and they provide a comfortable sense of enclosure by sealing off the unusable low space.

Framing the Walls

1 Marking Knee Wall Studs. Mark the height you want for the knee walls on a piece of 2x4 and, with a carpenter's level held against it to check for plumb, use the level to mark the rafters at the correct height. Measure the length of the knee walls and cut two 2x4 sole plates. Put the sole plates in position (but do not nail them down yet), set a stud on a sole plate resting against a rafter above at the mark for the wall height, and mark the angle of the rafter along the side of the stud. Cut and check the fit, then repeat the procedure along both knee walls. Also mark the sole plates for stud locations.

2 Adding Nailers. Turn a sole plate on edge and attach the short studs with 10d nails through the bottom of the plate. Stand the frame in position, check the fit, nail the sole plate to the floor with 16d nails driven into joists, and nail the studs to the rafters with 10d nails. Repeat the procedure for the other knee wall. Cut 2x4s to fit and attach on either side of the last stud on each sole plate.

3 Completing the Wall. Frame the center, squared section of the wall across the attic as you would a conventional stud wall, then raise and attach it to the collar beam. Nail cut-down stock to either side of the rafter from the collar beam to the last studs on the wall, attach the top plates on each side, and attach intermediate studs between the squared wall and the knee walls.

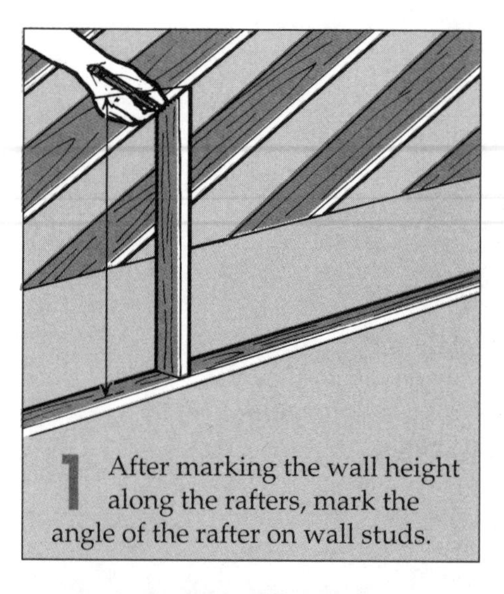

1 After marking the wall height along the rafters, mark the angle of the rafter on wall studs.

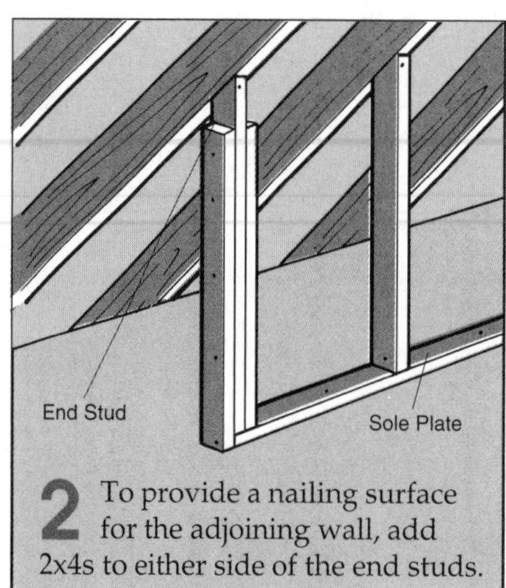

2 To provide a nailing surface for the adjoining wall, add 2x4s to either side of the end studs.

3 Widen rafters with registers between the collar beam and the knee wall to accept a top plate.

Installing a Collar Beam

Choose the rafter that you will build the wall upon. Measuring from the lower edge of the rafter, find and mark the point on the edge that is the height you want for the ceiling. Repeat the procedure on the corresponding rafter on the other side of the attic. Measure the distance between the sides of the roof at these points and cut two lengths of 2x6, 1/2 inch shorter than this distance. Find the angle of the roof and cut the ends of the 2x6s to conform. Nail short lengths of blocking the same thickness as the rafters between the 2x6s and install this doubled beam at the marks on the rafters; check that it is horizontal.

Framing Doorways

Making an opening for a door in a frame wall requires the elimination of a stud from the frame and the substitution of a short beam (called a header) across the opening and short studs (called cripples) between the header and the top plate.

Most do-it-yourselfers will choose to install a prehung door, because this factory-assembled unit entirely eliminates a very tricky bit of carpentry. Since the size of the opening obviously depends on the size of the door, purchase the door before starting the framing so you will know its exact dimensions.

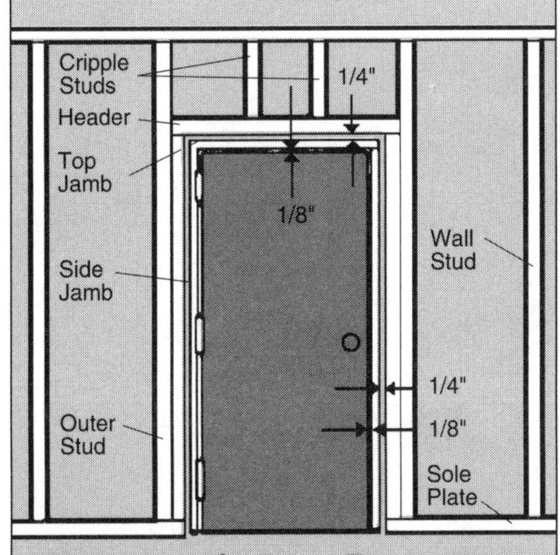

Anatomy of a Door Frame. The opening should be 1/2 in. wider and 1/4 in. taller than the outside dimensions of the jamb for adjustments.

Installing the Door

1 A Split Jamb Door. A prehung door for interior use comes with the jamb split; the door hangs on one jamb, the other jamb is inserted into the opening from the other side. Carefully unpack the assembly, but do not remove any wedges that hold the door closed.

2 Installing the Door. Insert the side of the jamb, with the door attached, into the opening. Shim the door closed if it is not already and support it at the bottom with shims. Adjust the jamb so that the door clears it by 1/8 inch all the way around, then nail the casing to the door frame with 8d finishing nails.

3 Shimming the Jamb. From the other side, fill the gap between the jamb and the jack studs and header with shims. Then nail the jamb through the shims to the frame along the sides with 16d nails. Fit the other half of the jamb into the opening, slipping the tongue at the top into the corresponding groove in the installed half. When it is seated, attach the jamb with 16d nails into the frame and nail the casing with 8d finishing nails.

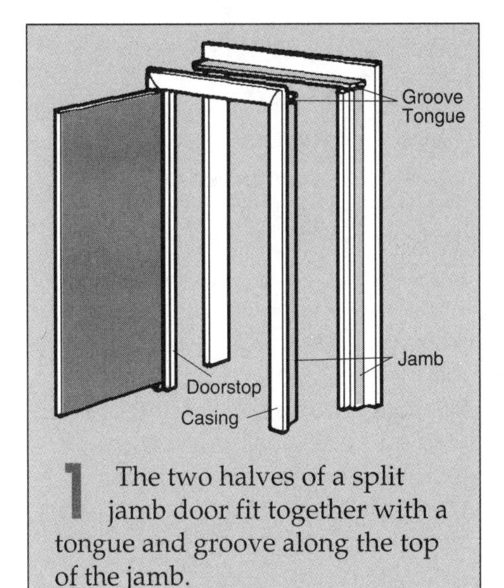

1 The two halves of a split jamb door fit together with a tongue and groove along the top of the jamb.

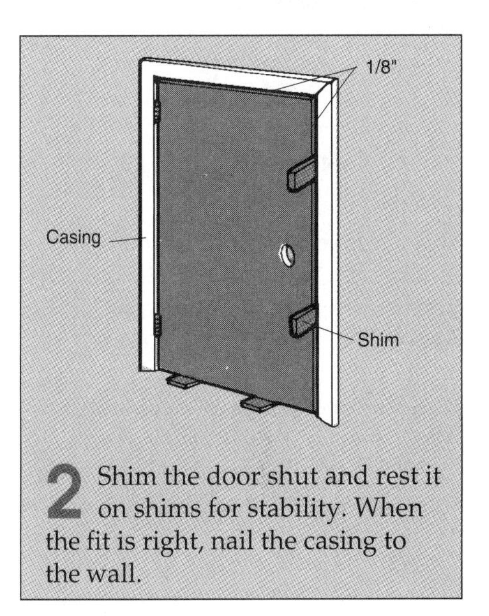

2 Shim the door shut and rest it on shims for stability. When the fit is right, nail the casing to the wall.

Framing the Opening

Use the dimensions of the door, plus the extra space and the width of the jack studs, to determine the proper location for outer studs on each side. Mark these on the top plate, and install using the other studs in the wall. Cut the jack studs to size and nail inside both outer studs with 1/2 dozen 16d nails. Cut the header to fit over the jack studs and attach with 16d nails driven through the outer studs. Attach short cripple studs above the header. Space the cripple studs so that the 16 or 24 inches on center sequence of studs along the wall is maintained.

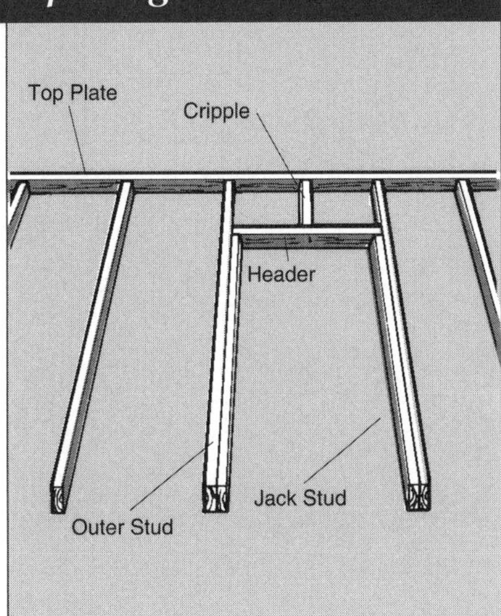

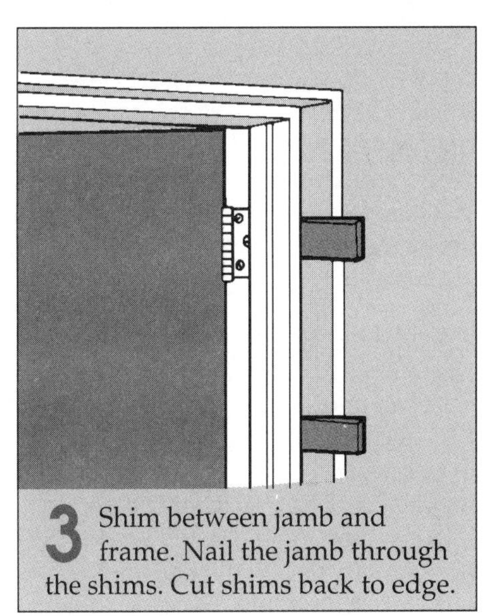

3 Shim between jamb and frame. Nail the jamb through the shims. Cut shims back to edge.

Framing with Metal Studs

Lightweight metal framing is often easier to install than wood framing and provides a wall that is just as sturdy. If you are concerned about extra soundproofing, wood should probably be your choice, but for privacy with moderate resistance to sound, a metal frame wall is an alternative worth considering. Some building codes, especially those governing city loft conversions, require metal framing for fire protection. Metal framing consists of studs and the tracks they fit into. They both are long, U-shaped members. Studs are the heavier of the two, with the flange at right angles to the spine. Tracks, lighter in weight, have flanges angled slightly toward the spine to grip studs when they are snapped in place.

Metal framing can be cut to fit with straight tin shears. It can be purchased in kit form for small jobs and by the piece, in any length, for larger projects. Metal frames are put together and wallboard is attached with metal screws. You will need a variable speed power drill or a power screwdriver to make the joints tight.

Putting Up a Metal Stud Wall

1 Cutting the Studs. Plan for metal framing just as you would for wood framing, but use wood as the inside framing in doorways that are framed with metal (see page 23). All necessary corners can be made with simple cuts. Remember to plan for extra studs at corners for attaching wallboard.

2 Making Corners. Cut the metal tracks to length with tin shears. You can attach the bottom track with adhesive, or you can do so with nails or screws. When using adhesive, make sure the floor is level, clean, and provides a sound surface for the adhesive to grip (not loose linoleum or a floor with gaps). Run a 3/8-inch bead of panel adhesive in a 1-inch-wide squiggle along the chalk line you have snapped for the track. Seat the track on the line and press firmly into the adhesive; allow to dry according to manufacturer's instructions. To attach without adhesive, use masonry nails for a concrete floor, or spiral nails or sheet metal screws for wood.

3 Seating the Studs. Use a plumb bob to mark the position for the top track and install it across joists, along one joist, or across nailers in an exposed ceiling—or use toggle bolts (see page 52) to attach the top track to a finished ceiling. Cut the metal studs to length if necessary and insert them sideways into the tracks. Snap them into place and check to be sure they are plumb. The tension applied by the flanges of the plate should hold the studs in place, but for extra strength attach them with sheet metal screws.

4 Attaching Wallboard. The frame will not be completely rigid until the wall surface is installed. When you attach wallboard to the metal studs, use a portable drill or power screwdriver to drive drywall screws (see page 28).

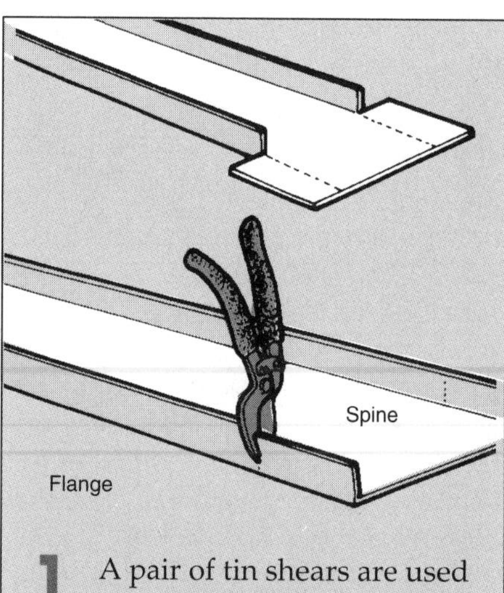

1 A pair of tin shears are used to cut the studs and tracks of aluminum framing to fit.

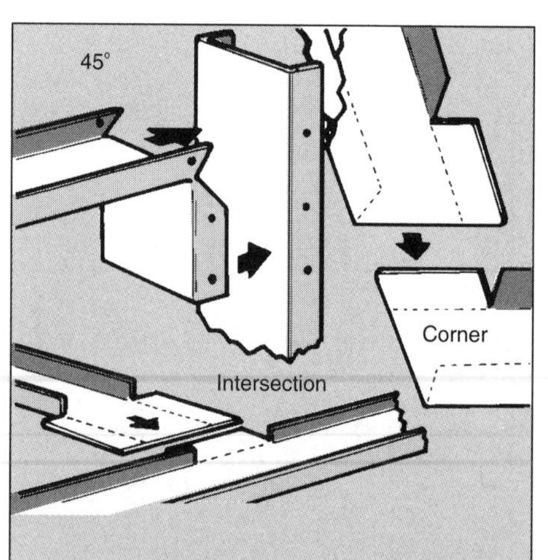

2 Intersections and corners are made by cutting the flanges, then overlapping the studs as indicated.

3 With the track in place and the studs cut to length, set one in place and twist until it snaps in tightly.

4 The most efficient way to attach wallboard to metal studs is with a power screwdriver and wallboard screws.

Installing Wall Outlets

This page contains some basic information that you should keep in mind when removing or adding walls. For extensive information on electrical wiring and installation, consult an electrician or refer to other, more detailed, texts on the subject.

The absolute first rule for working with electricity is to turn it off. The second rule is to make sure that you have turned it off. Electricity is dangerous, but wiring and electrical fixtures that have been shut off at the source of power—your fuse box or circuit breaker—are harmless.

To shut off power to part of your house, you must remove the fuse or flip the circuit breaker that controls power to that line. If you do not know what circuit a fuse or circuit breaker affects, turn on all lights controlled by switches in one area, and plug lamps into all the outlets in a room and turn them on. Station a helper in the room to call out when the lights go out. Use all the outlets in this test because different sides of a room may be on different lines. If you are not sure whether there are live lines running to other parts of the house through a wall you want to remove, you can be sure to avoid hazard by shutting down all power in the house. If you encounter wiring routed through a wall to other areas, you should consult an electrician.

Wall Boxes for Outlets & Switches

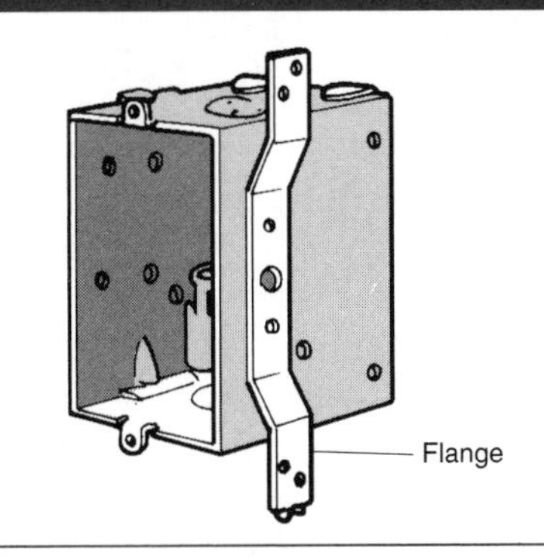

Outlets and switches are mounted in metal boxes attached to studs in the framing. The most common kind has a flange on one side that is nailed to the stud. The box can be mounted with either end up so it can be oriented toward either side of a frame wall and on either side of a stud.

Mounting Wall Boxes

When you mount wall boxes to a frame wall, be sure to account for the thickness of the surface that will cover the wall. The box should protrude from the frame a distance equal to the depth of the wallcovering.

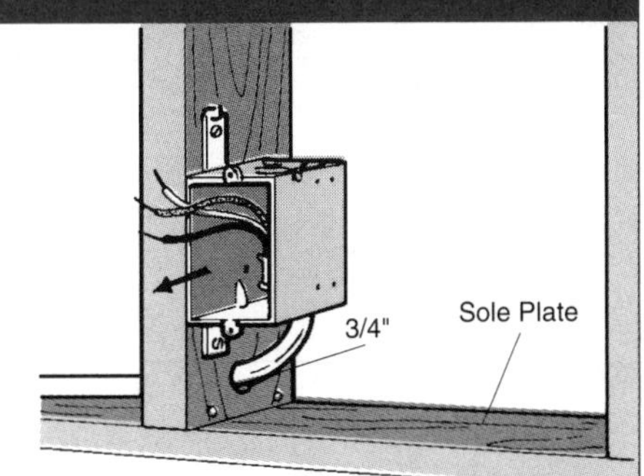

Running Electrical Cable Through Framing

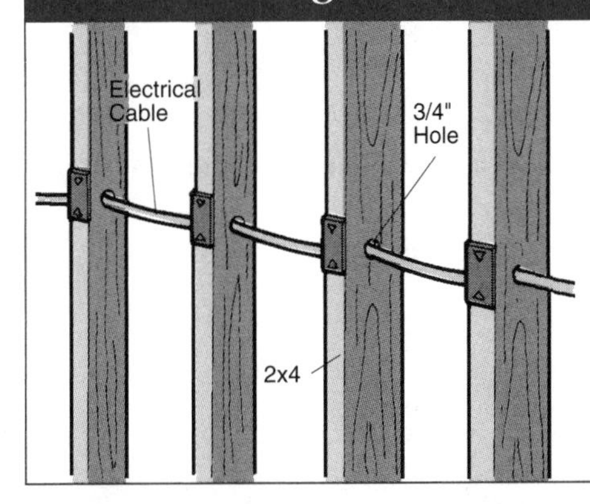

To run cable through a stud wall, drill 3/4-inch holes through the middle of the studs. The National Electrical Code requires that you put metal plates (available at electrical supply houses) on the edge (or edges) of the stud that faces out to guard against the possibility of later driving a nail through the stud into the cable.

Testing for Current

A voltage tester is an inexpensive device that lets you know when current is present in a line. It consists of a small light bulb and two probes that will fit into the slots in a plug. Use the tester for both plugs and switches, touching one probe to the black wire in a pair of wires, the other probe to the white wire on the other side of the switch or outlet; test all sets of wires in an outlet or switch. When current is present, the bulb will light. You can use a lamp for a voltage tester simply by plugging it in and turning it on, but first make absolutely sure, for obvious reasons, that the bulb in the lamp is not burned out.

Wallboard

Wallboard is used for the surface of most interior walls built today. Also known as plasterboard, drywall, gypsum board, or by the trade name Sheetrock, wallboard is easy to handle, can be installed quickly, and costs relatively little. It provides a surface that can be paneled or tiled as it is, or finished with joint compound and tape and then painted or wallpapered.

Wallboard is plaster sandwiched between layers of strong paper. It is fire resistant and stops sound well. Regular wallboard has a dark gray kraft paper backing and smooth, off-white paper on the front that takes paint readily. The long edges are sealed and slightly tapered to accept tape and compound.

Some types of wallboard have special purposes. Water-resistant wallboard, with water-repellent cores and facing, is usually tinted blue or green. It is made for use in areas of high moisture or as a base for ceramic tile. Type X wallboard with extra fire resistance, is sometimes required by building codes for special places, such as on walls between an attached garage and the main living area. Insulating wallboard has an aluminum foil backing to be used on the inside surfaces of exterior walls. Backer boards are used under another layer of wallboard or paneling. Pre-finished wallboard is faced on one side with a durable finish that needs no paint.

Wallboard is available in different thicknesses and panel sizes. Thicknesses are 1/4 inch, 3/8 inch, 1/2 inch and 5/8 inch. Half-inch wallboard is used for most walls and ceilings, while 5/8-inch wallboard is used for extra soundproofing; 1/4-inch wallboard is used for putting a new surface over an existing one and 3/8-inch wallboard is used to multiply soundproof walls and ceilings below attics. All panels are 4 feet wide and range in length from 6 to 16 feet for the two thinner versions, to 12 feet for 1/2 inch, and 8 feet for 5/8-inch thickness.

Hanging a wallboard surface on a wall is easy if you observe a few basic cautions. Most importantly, remember that wallboard is delicate. It crushes if stood on one corner (a 4x8-foot panel weighs 65 pounds) and breaks easily if dropped or hit. Treat it relatively gently. Wallboard should be stacked flat to prevent warping if not used right away.

To estimate the materials you need for a job, find the number of square feet you will cover by multiplying the outside dimensions of the wall (or walls), subtracting the area of any openings, then adding 10 percent of the total for wastage. Divide by the number of square feet in the panel size you plan to use and the result is the number of panels to order. Wallboard is fastened to studs with 1/2-inch, ringed wallboard nails.

Cutting Wallboard

1. It is often necessary to cut down the length or width of a wallboard panel. For long, straight cuts of this kind, use a utility knife and a straightedge. Put the wallboard on the floor or on the stack of wallboard you are installing; then measure and mark the cutting line. Use a carpenter's square as a straightedge to hold the knife on the line and score the paper, cutting through it into the core.

2. If you are working on a stack of wallboard, slip the cut over the edge of the stack and snap the panel along the cut. It should break neatly with the two parts held together only by the paper backing. If you are working on the floor, slip a piece of scrap under the cut and snap it.

3. Turn the wallboard on edge, fold the two cut pieces slightly together, and slice through the paper backing to complete the cut. The resulting edge will be somewhat rough. Cut panels 1/4 inch less than measured dimensions to account for the roughness.

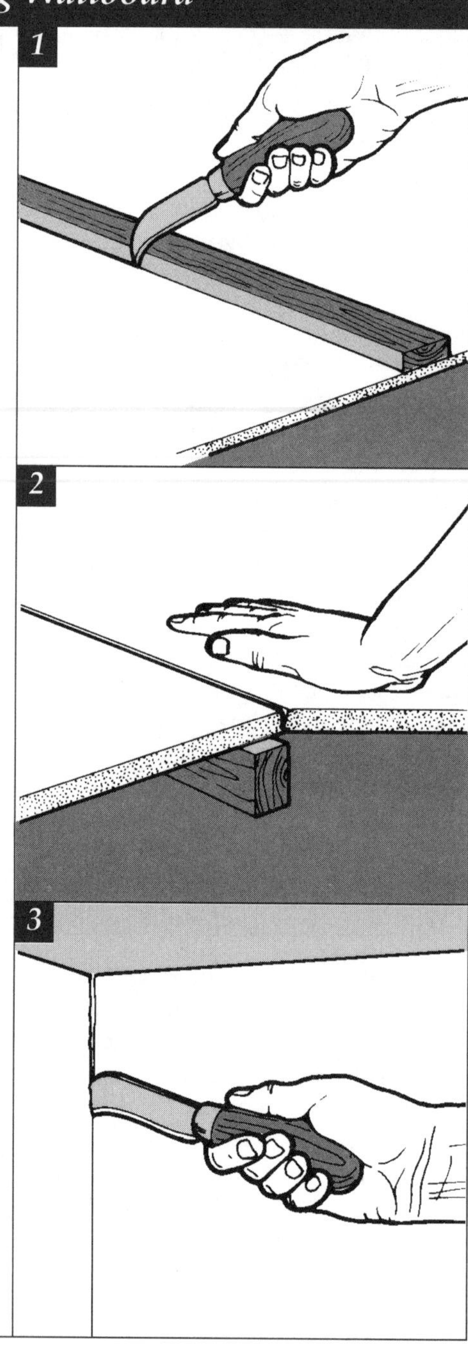

Cutting Openings in Wallboard

1 Marking the Cut. You will have to cut holes in some pieces of wallboard as openings for outlet boxes and switches on a wall. You also may have to cut away a section of wallboard to fit a panel around a doorway. To cut an opening, first mark the position.

2 Marking an Outline. You can mark the entire outline of the opening from measurements, or if you want a more precise cut, use the object the hole must fit around as a template for drawing the outline. In either case, enlarge the outline 1/8 inch all the way around to ensure a good fit.

3 Starting the Cut. Punch through the wallboard with an awl or a nail at each corner of the outline so that these points will show on the back. Score the outline with the utility knife so that it cuts 1/4 inch or a little more into the core.

4 Marking the Cut. Flip the board over and mark the outline between the corner holes that show through from the front. Score the board along these lines as you did on the other side.

5 Completing the Cut. Turn the board to the front side and tap out the scored piece with a hammer. The edges of the cut can be cleaned up or enlarged by carving with the utility knife. To cut an indent to fit around a door or window, mark off the area and score all but the longest line as above, but in this case cut all the way through both sides of the wallboard. Finally, score the longest cut and snap off the piece as shown on the previous page.

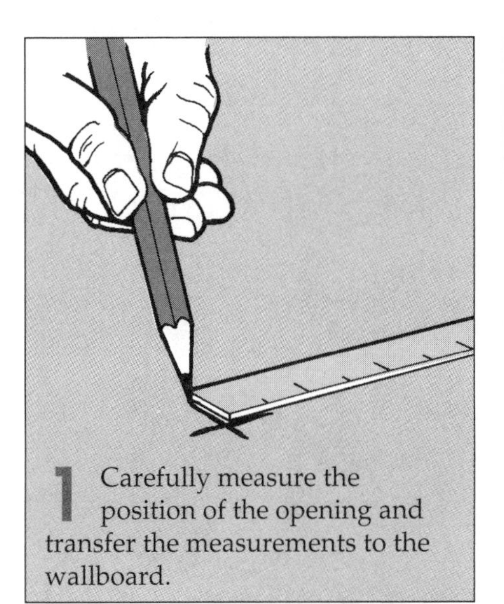

1 Carefully measure the position of the opening and transfer the measurements to the wallboard.

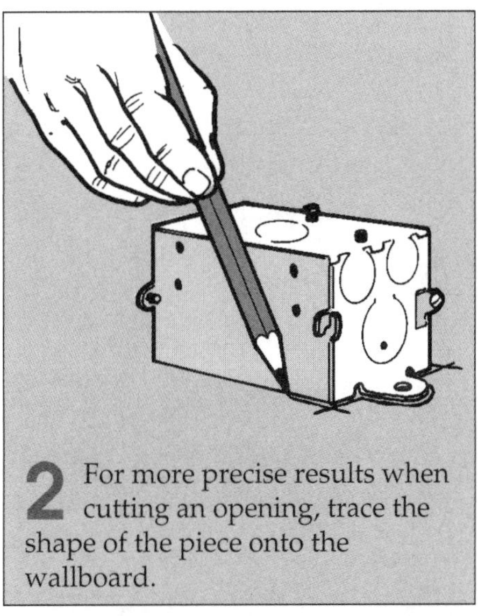

2 For more precise results when cutting an opening, trace the shape of the piece onto the wallboard.

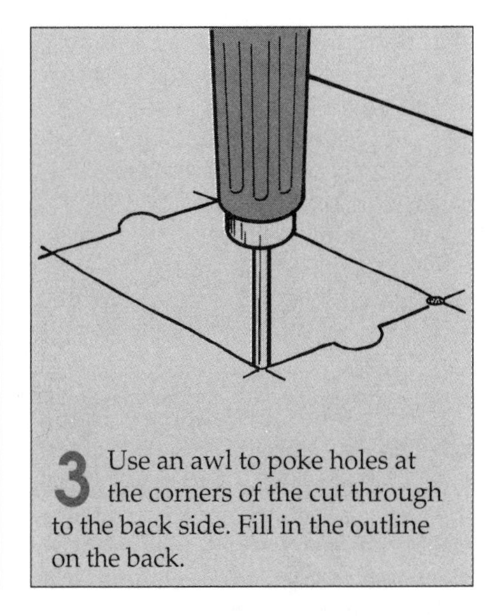

3 Use an awl to poke holes at the corners of the cut through to the back side. Fill in the outline on the back.

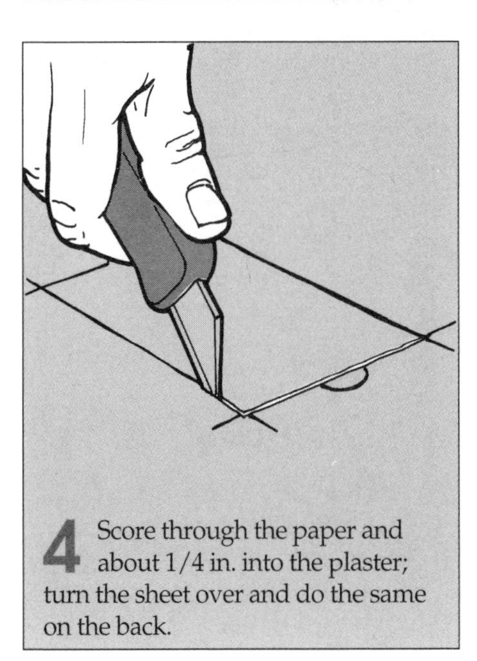

4 Score through the paper and about 1/4 in. into the plaster; turn the sheet over and do the same on the back.

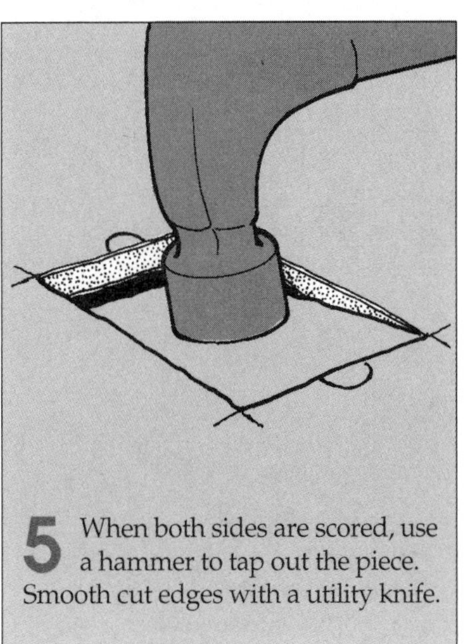

5 When both sides are scored, use a hammer to tap out the piece. Smooth cut edges with a utility knife.

Cutting with a Saw

You can cut small openings in wallboard with a keyhole saw by drilling out opposite corners of the outline to start the cuts. Any indents or other odd shapes can be finished with a utility knife.

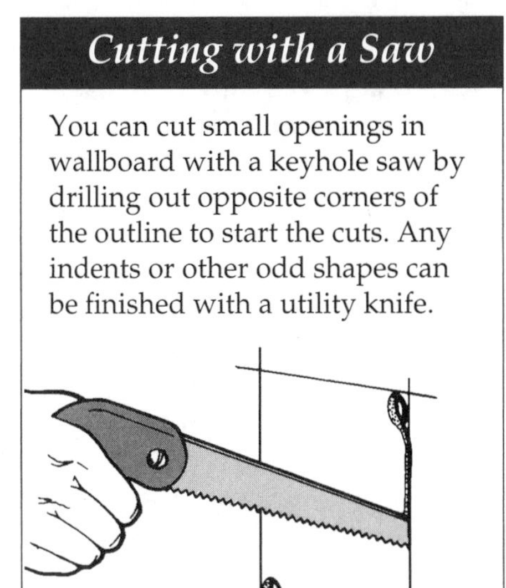

Installing Wallboard

1 Planning the Job. Check the stud wall frame to be sure that it provides nailing surfaces for all edges of the wallboard and add studs where needed. Also check that outlets and switches are positioned correctly (see page 25). Mark the exact location of the wall studs along the floor and ceiling, so you will know where to nail. You may use wallboard nails (procedure described below) or drywall screws. For easier handling, cut wallboard about 3/4 inch shorter than the height of the wall. The gap will go at the bottom to be covered with base molding.

2 Setting Wallboard in Position. Although horizontal installation of wallboard can reduce the number of joints to plaster (see next page), most do-it-yourselfers find it more convenient to hang panels vertically. To position a panel for attachment, make a lever out of scrap wood and carefully stand the panel in position against the wall on the end of the lever. Apply pressure with your foot.

3 Securing the Wallboard. Hold the panel against the ceiling with the lever and push it tightly against the wall with your hand. Drive a few wallboard nails through the panel into the frame at the top and along a side. This will hold the panel in place and allow you to complete the nailing.

4 Nailing the Board. Drive nails slightly below the surface of the wallboard but be careful not to break the paper. Dimples left by countersinking will be filled with compound.

5 Completing the Nailing. Standard procedure calls for nails at intervals of 6 inches along all wallboard edges and 12 inches along studs in the middle. Nails around the perimeter should be driven in 3/8 inch from the edge of the wallboard. If a nail misses a stud, just pull it out and renail. For added protection against nails popping later, use pairs of nails 2 inches apart at the same intervals.

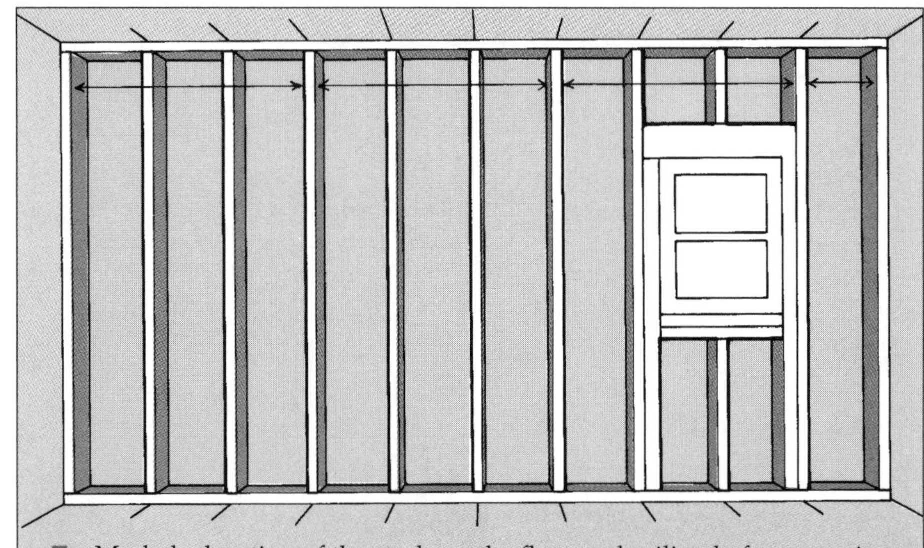

1 Mark the location of the studs on the floor and ceiling before covering them with wallboard so nailing will be easy when the studs can't be seen. Plan whatever cuts you must make and be sure that you have a nailing surface wherever pieces meet.

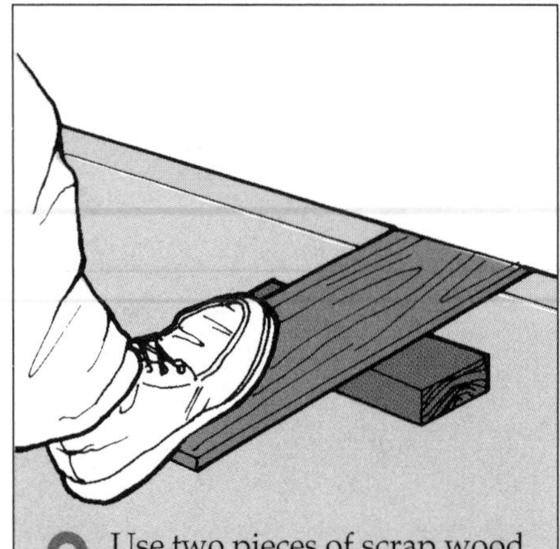

2 Use two pieces of scrap wood to lever the wallboard snugly against the ceiling.

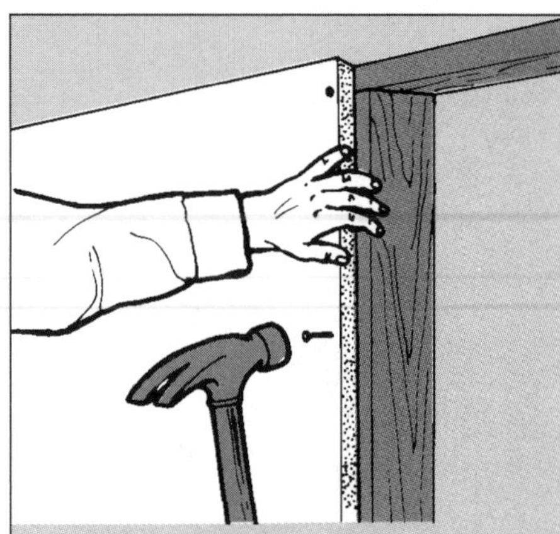

3 With the sheet held against the ceiling by the lever, drive nails into top and along one edge.

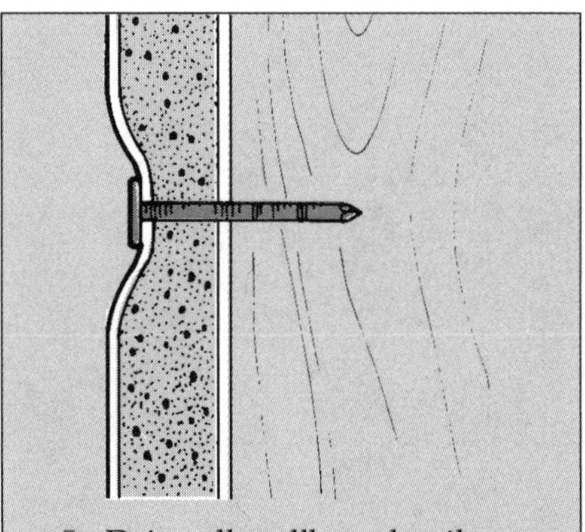

4 Drive all wallboard nails about 1/32 in. below the surface—the paper will dimple. Take care not to break the paper.

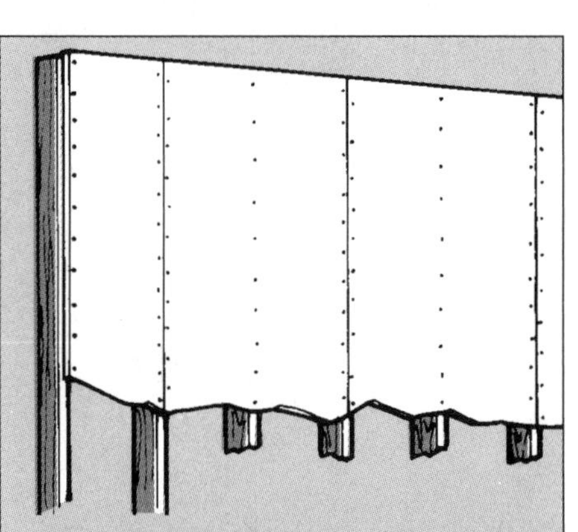

5 Drive wallboard nails at 6-in. intervals along all the edges and into center studs at 12-in. intervals.

Installing Wallboard with Adhesive

1. You can reduce the amount of nailing by using an adhesive to attach wallboard to the studs behind the middle of the panel. Use a panel adhesive (check with your lumber yard for the right one to use with your wallboard) applied to the center studs in a 3/8-inch squiggle.

2. Start and stop the application of adhesive 6 inches from the top and bottom of the studs. Fit the panel against the studs and nail the perimeter. If the wallboard bulges away from a center stud, press it back against the stud and nail it down. One way to improve adhesion against studs is first to let the sheets of wallboard warp slightly. Set them individually face up on 2x4s at each side and allow them to sag for a day. The resulting curvature will press against the studs when the wallboard is nailed top and bottom.

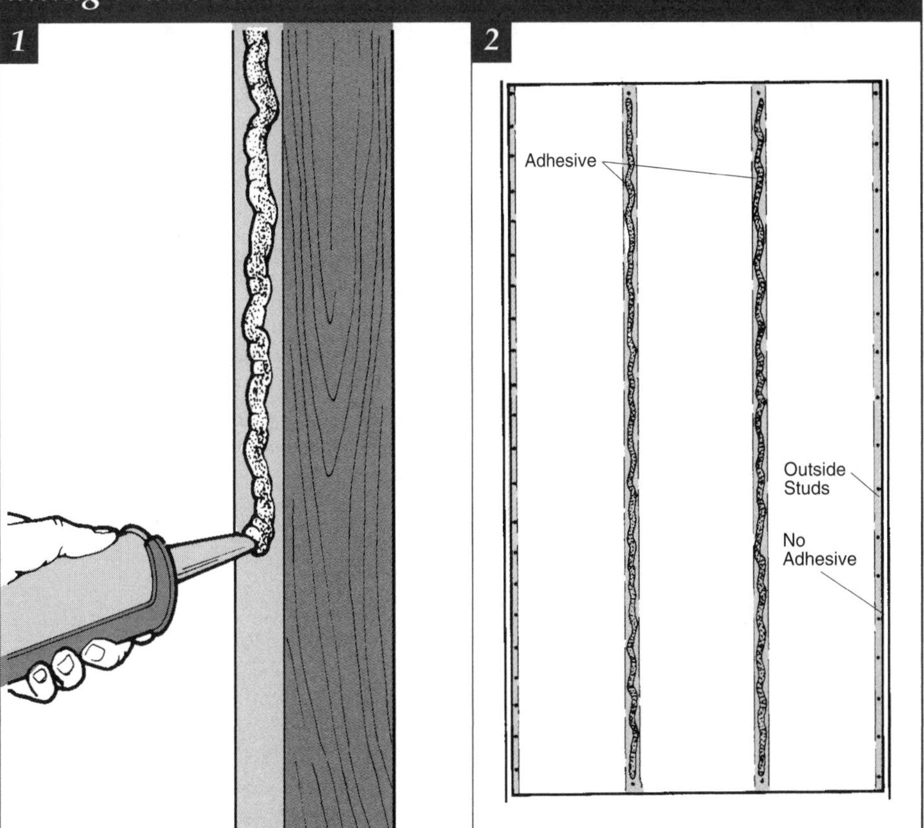

Hanging Wallboard Horizontally

If you have help, you can save on the number of feet of wallboard seams you have to finish—and improve the look of the finished wall—by hanging longer panels horizontally.

To hang wallboard horizontally, mark the stud locations on the floor and ceiling as in the previous procedure and drive 8d or 10d nails halfway into the studs 4 feet down from the ceiling. With a helper, lift the panel into position onto the nails and push back against the wall. Nail the panel to a few of the studs to hold it and complete nailing as in a vertical installation. If the wall is 8 feet high, install the panel below using a lever to hold it snugly against the one above. If the wall is higher than 8 feet, repeat the procedure given here and fill in between the lower panel and the floor with pieces cut to fit. Measure the distance between the floor and the bottom panel in several places in case the floor and ceiling are not square.

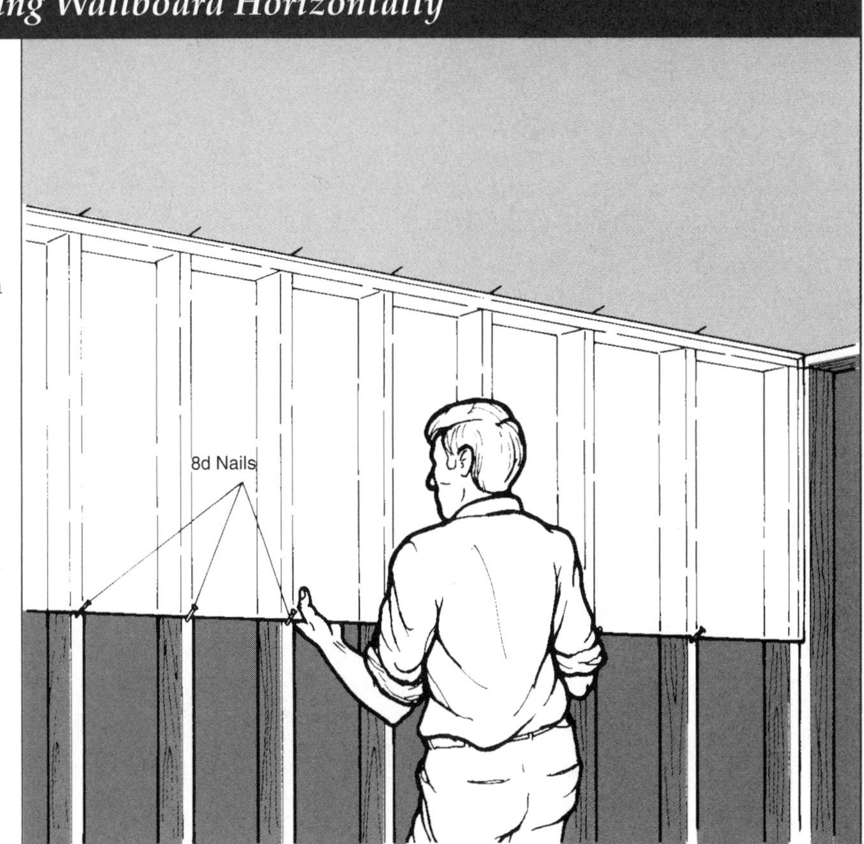

Taping & Feathering Wallboard

A thick spreadable substance called joint compound, a few rolls of perforated paper or fiberglass mesh tape, and some patient labor will turn a newly hung wallboard wall, with its unsightly seams and nails showing, into a uniformly smooth surface ready for paint or wallcovering.

First, the seams between the wallboard are taped. A finishing knife is used to apply compound over the tape, and then with successive strokes, the edge of the blade is used to taper the compound. This technique is called "feathering."

Taping and feathering will hide a multitude of missed hammer strokes and badly fitting joints, but nothing (except a new surface) can hide a poor job of taping and feathering. Remember, when taping and feathering that your work will show.

The nature of joint compound enforces patience: It must be applied at least three times and left to dry about 24 hours after each application. (There are special fast-drying joint compounds but they are not recommended for anyone with less than expert skills.) Joint compound can be purchased premixed or dry. The dry type has to be mixed at home. Unless you are experienced with mixing your own, use the premixed kind—it costs only a little more and is much easier to use. Premixed compound is easy to store. Wipe clean all exposed surfaces inside the can (so bits will not dry and flake off into the mixture) and reseal. If you are storing compound for more than a couple of days, cover the surface with 1/2 inch of water to seal it, then pour it off before using again. Be sure not to store compound where it might freeze.

The 2-inch-wide paper tape that fills the seams between wallboard panels is available in 250-foot rolls. There are two basic types: one has holes that are so small they are barely visible; the other is a fiberglass mesh tape which is self adhesive.

The tools used to apply tape and joint compound are simple: a hawk (see left) to hold the compound, 4-inch and 10-inch finishing knives for spreading it, and 100-grit, open-coat sandpaper to smooth it when dry.

A Step-Saving Tool

A hawk is a hand-held workbench for feathering. It holds a supply of compound where you need it and provides an edge for scraping compound off the finishing knife. You can repair a wall without one, but you will have to bend to the can of compound hundreds of times to load and clean your knife.

Taping & Feathering Flat Seams

1 Filling Nail Dimples. Inspect the wall closely, checking that each nail is set below the surface. Any nails protruding or flush with the surface should be recessed about 1/32 inch. Scoop up a bit of compound at the end of the 4-inch knife and fill a nail dimple by spreading compound over it with the knife. Clean away any excess. Filling the nail dimples first is a good way to get the feel of working with joint compound. Sand and apply additional finishing coats as you finish the seams, using the procedures given in the directions below.

Anatomy of a Taped Seam

The basic sequence of taping and feathering a seam is shown in the drawing below. First a layer of compound is spread down the seam, then the length of the seam is covered with tape. The tape is then imbedded in the compound with a second pass of the knife. This provides the filling over which finishing (feathered) layers are added to make a smooth seam.

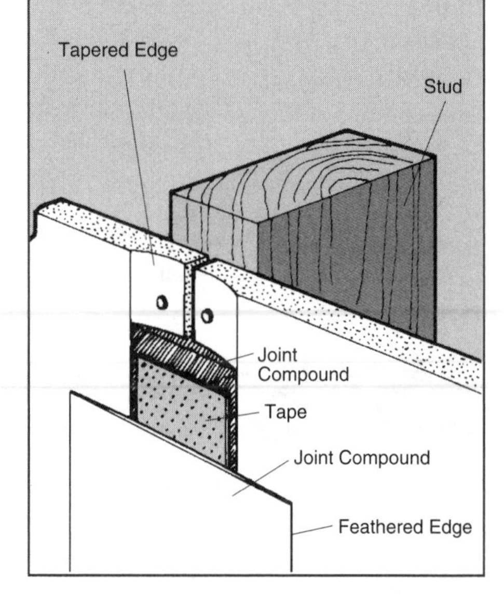

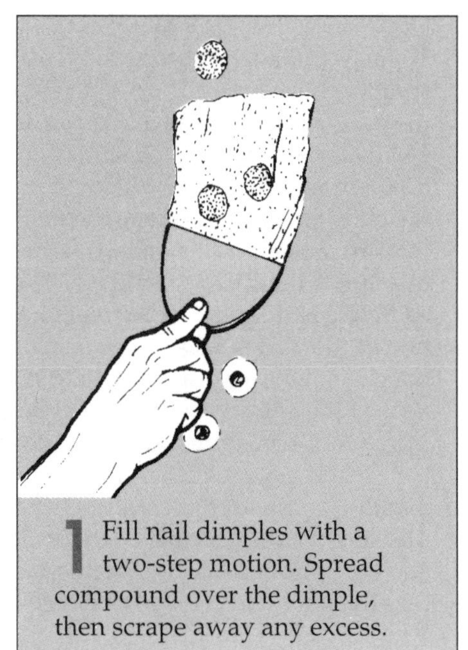

1 Fill nail dimples with a two-step motion. Spread compound over the dimple, then scrape away any excess.

2 Starting a Seam. Work on one seam at a time, repeating the procedures in steps 2 to 5 for each. Pick up a supply of compound on the end of the 4-inch knife and spread a layer about 1/8 inch thick down a seam with the knife at a 45-degree angle. Fill the seam from top to bottom, making sure that you leave no gaps, which will become bubbles, under the tape. This application need not be neat, but it must be thorough.

3 Fixing the Tape. Tear off a piece of tape the length of the seam. Starting at the top, lay the tape carefully over the center of the seam so that it runs straight down. Anchor the top of the tape with some compound. Keep the tape straight and smooth.

4 Imbedding the Tape. Work the tape down from the top, pressing the blade firmly against the joint to imbed the tape. The pressure on the knife should stretch the tape slightly, but do not to tear it. If it does tear, discard the tape below the tear and replace with new tape, allowing the ends to meet but not to overlap. If the tape wrinkles, lift it up and lay it back down flat, then go over it again. Eliminate any air bubbles in the tape.

5 Removing Excess Compound. Go back to the seam and scrape away any excess compound that has gathered at the edges. Clean the knife on the edge of the hawk. Then apply a very light coat of compound over the taped seam.

6 Sanding the First Coats. Allow the compound to dry for at least 24 hours. Sand down any rough spots and raised areas with 100-grit, open-grain sandpaper. Wear a filter mask when sanding.

7 Applying Finishing Coats. Use a 10-inch knife to apply at least three coats, allowing each one to dry thoroughly. The applications should be successively wider by a few inches on each side with the edges feathered out (or blended) into the bare wallboard. Check the final finish coat for rough spots (see page 33).

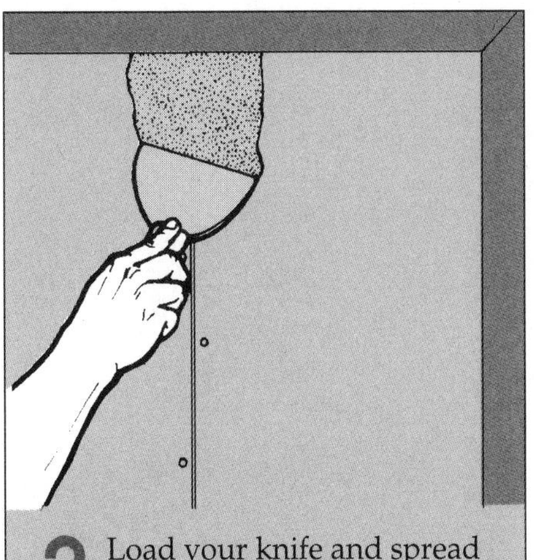

2 Load your knife and spread compound about 1/8-in. thick down the open seam from top to bottom.

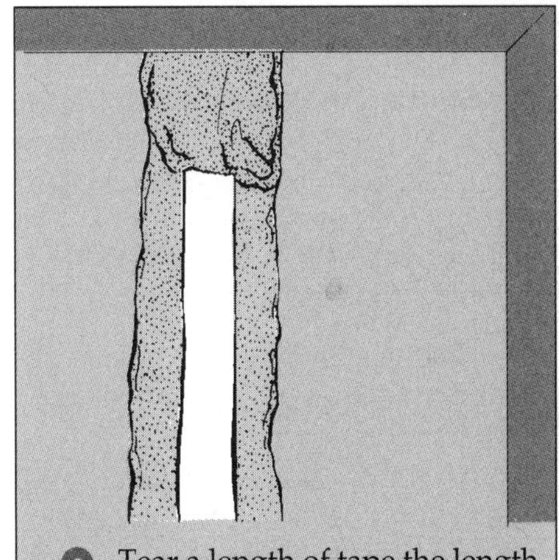

3 Tear a length of tape the length of the seam and fix it at the top with a layer of compound to hold it in place.

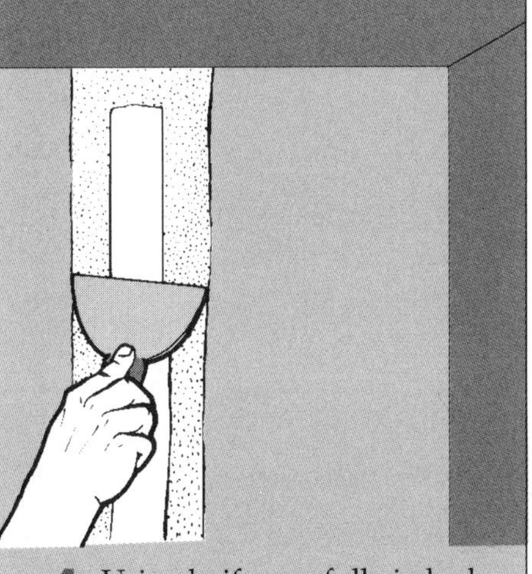

4 Using knife, carefully imbed the tape in the compound; lift tape to take out wrinkles.

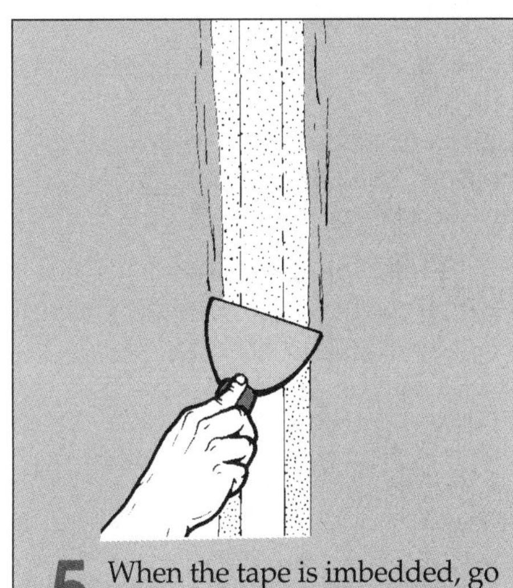

5 When the tape is imbedded, go back to the top of the seam and scrape away any excess.

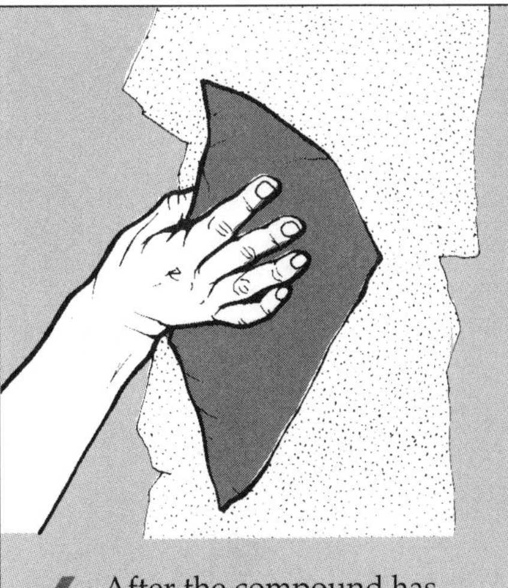

6 After the compound has dried at least a day, smooth down any rough spots.

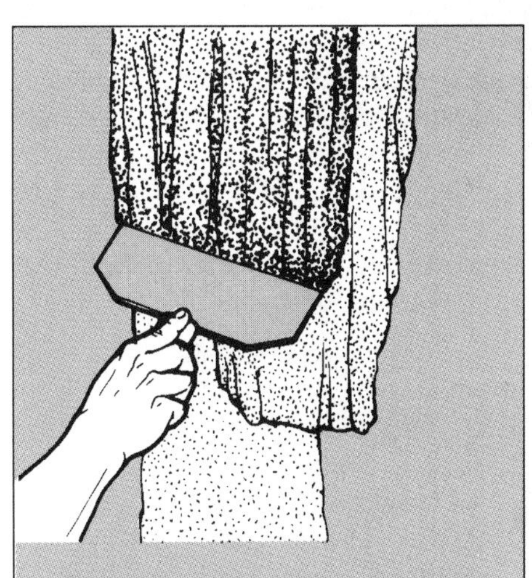

7 Use a wide knife to apply thin finishing coats of compound to seam. Several coats are required.

Finishing Inside Corners

1 Starting the Corner. Apply a first layer of compound down both sides of an inside corner. Make sure that the compound fills the corner itself all the way from top to bottom so tape will adhere.

2 Folding the Tape. Tear off a piece of tape the length of the corner and fold it down the center lengthways. Ordinary wallboard tape is pre-creased down the center so it folds easily.

3 Applying the Tape. Carefully lay the folded tape into the corner and, starting from the top, press it into the compound. Then gently poke the crease into the corner with your knife blade. Smooth the sides of the tape onto the adjacent walls with the knife, taking care not to pull the tape away from the corner.

4 Finishing the Corner. Apply a second layer of compound over the tape, down one side then the other, taking care not to pull the tape out of the corner with your strokes. Finish with several more applications of compound, sanding between, as with flat joints.

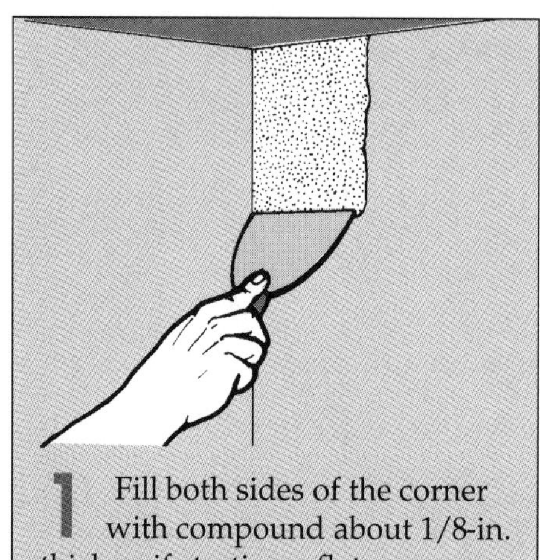

1 Fill both sides of the corner with compound about 1/8-in. thick as if starting a flat seam.

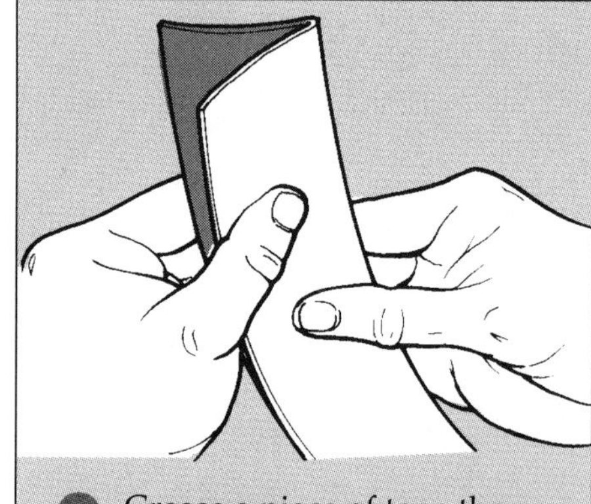

2 Crease a piece of tape the length of the corner down the center so that it will fit into the corner.

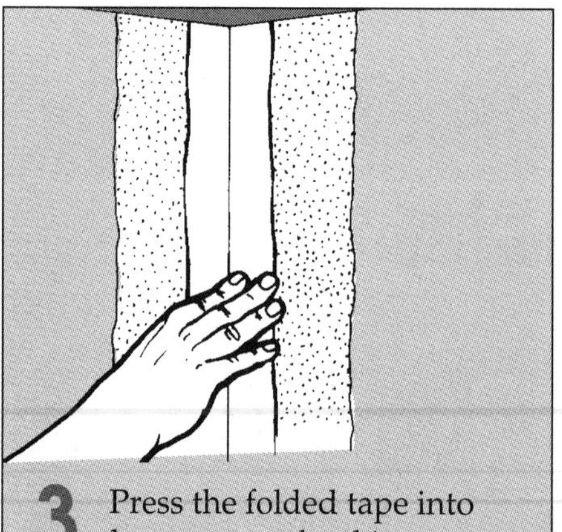

3 Press the folded tape into the compound, taking care that the crease fits exactly into the corner.

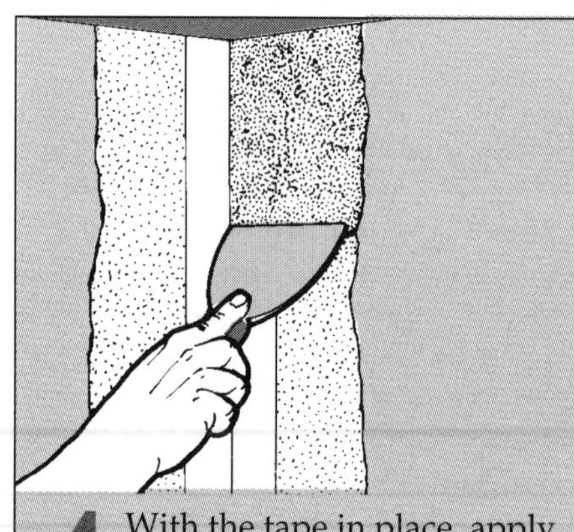

4 With the tape in place, apply a layer of compound over it, scrape away excess and finish.

Finishing Outside Corners

1. Outside corners need extra protection from damage. This is provided by a lightweight right angle bracket called a corner bead. Cut the bead to length with tin shears and nail it onto the corner, but do not try to dimple the nails. The corner itself sticks out slightly from the wall and compound will cover nail heads.

2. Apply a layer of joint compound by loading the knife and working from the wall out to the corner with horizontal strokes. Then scrape off the excess, working down the corner with vertical strokes. The blade should scrape the wall on one side and overhang the corner on the other. Allow to dry and finish with several more coats.

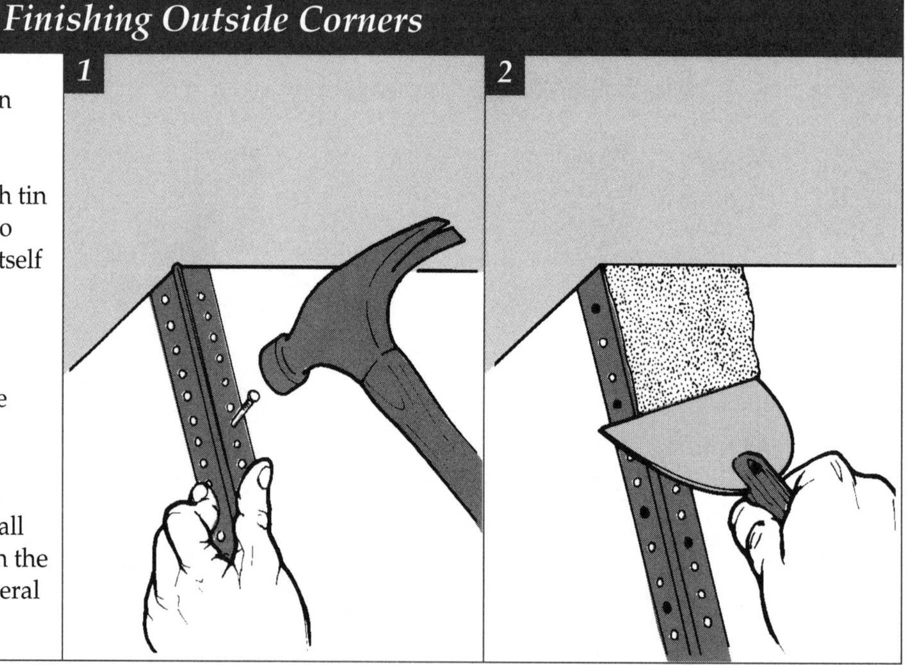

Checking the Finish

To make sure the job is perfect when you are finished applying the coats of compound, use a light to check your work—it may show problem areas you would not otherwise see. Shine the light along the wall (this works best if the light is directed by a hood, like a photo flood) and look for raised areas and depressions revealed by shadows. Smooth such raised areas with sandpaper and fill depressions with more compound.

Eliminating Bubbles

1 Removing a Bubble. If the tape has separated from the compound at any point, it will rise as a bubble when the compound is dry (or even later, when the wall has been painted or covered). Check the seams closely for bubbles when the job is finished. Any you find can be corrected quite easily. Use a utility knife to cut around the bubbled area just into the tape that is sound.

2 Refinishing a Bubbled Spot. Remove the bubbled tape from the wall and fill the resulting depression with compound as if it were a nail dimple. Clean away excess, allow to dry, and repeat applications of compound, sanding after each one, until the surface is smooth.

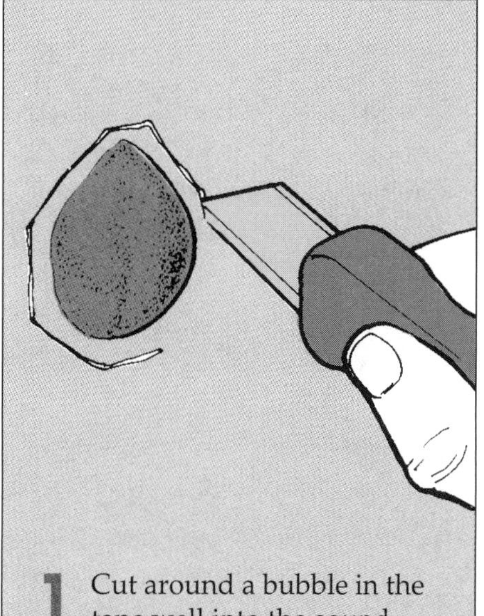

1 Cut around a bubble in the tape well into the sound surface, so that no air gap will remain when you peel off tape.

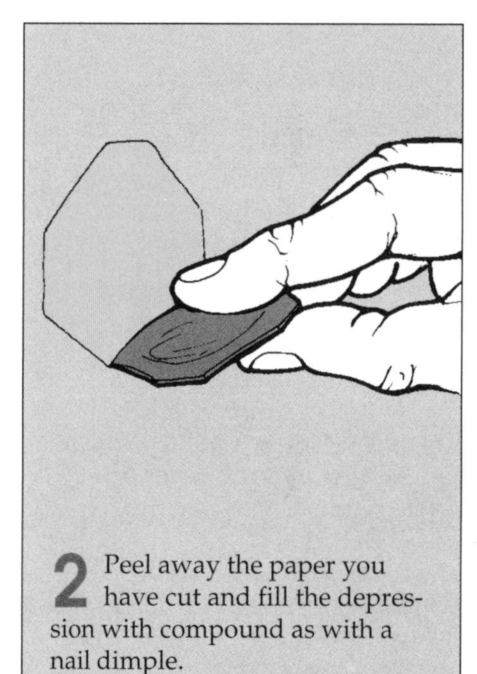

2 Peel away the paper you have cut and fill the depression with compound as with a nail dimple.

Filling Gaps

1. If a hole that was cut into the wallboard to fit around an outlet or switch is slightly too large for the face plate to cover, it can be built up with compound and tape. Apply a generous amount of compound over the edge of the gap, so that it is filled.

2. Lay a piece of tape over the compound with one long edge against the edge of the electrical box. Anchor the tape with another layer of compound over it, then scrape away the excess, taking care not to pull the tape out of position as you do so. Finish with several more layers.

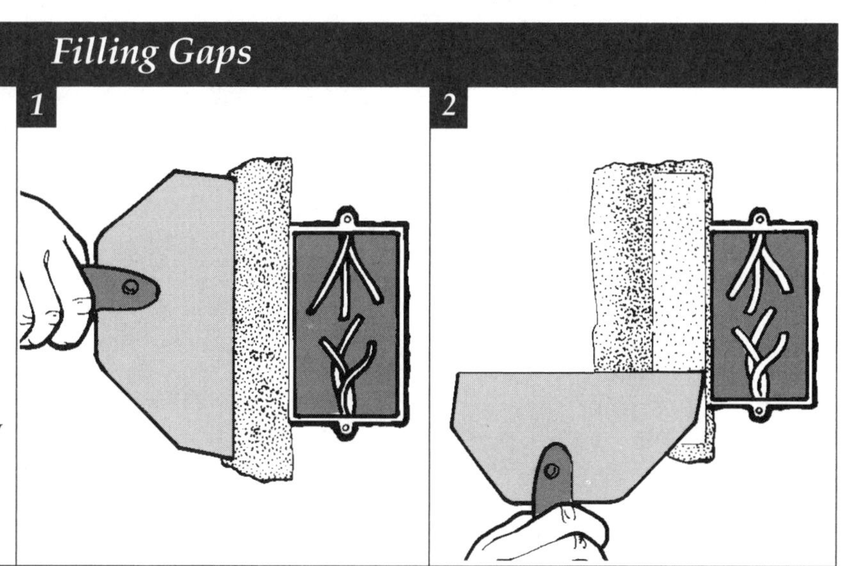

Installing Sheet Paneling

Sheet paneling—of real or simulated wood—is one of the most popular wall surfaces for do-it-yourself installation. It is one of the easiest of all wall surfaces to install and it quickly transforms a space from ordinary to special.

All paneling is sold in 4x8-foot sheets; although there are larger sheets available at a premium price in some grades. Sheet paneling is a veneer of real wood, vinyl, paper or non-descript wood printed to resemble fine wood grain. The veneer is laid over a composition wood. It is usually grooved vertically to imitate planks of different widths. Grooves fall in 16- and 24-inch intervals to hide the nails that hold the sheets to the furring or studs.

There are four basic types of paneling: Wood veneer on plywood —as the name suggests—is real wood veneer (anything from pine to exotic woods) over plywood. It is usually 1/4-inch thick and is finished with a clear plastic coating that protects the wood from damage. Because the veneer is real, each panel is unique. This is the strongest and most expensive type of paneling; and prices vary based on the type of veneer used. Simulated veneer on plywood is a panel covered on the good side by a layer of paper, vinyl, or wood printed to resemble a fine wood veneer. It is 5/32 or 3/16 inch thick and somewhat less expensive than real wood veneer on plywood. Unlike real wood, these panels repeat the surface pattern. Even less expensive is patterned hardboard, a simulated surface (fabric patterns as well as wood) on hardboard, either 5/32 or 3/16 inch thick. Hardboard paneling is not as strong as either kind of plywood paneling, but it is less costly. Finally, the least expensive type of paneling is patterned particle-board, a printed paper, vinyl or wood layer on 3/16-inch particleboard. It is rigid, somewhat brittle and not as easy to install as the other types of sheet paneling, but it is useful where economy is important.

There are a few basic rules for purchasing paneling that will improve the look of your job. First, always plan the work carefully before ordering the paneling so that you do not buy too much or too little. When you choose the paneling, try to buy it from stock so that you can inspect each sheet for undamaged edges. Find out whether you can get molding that will complement the paneling you choose and consider using the specially colored panel nails that are available to blend with panel colors.

Planning the Job

Make a scale drawing of the room you plan to panel. This is a handy reference for determining where it is best to locate the first sheet of paneling, where you need additional furring, and where you will have to make tricky cuts. For ordering materials, divide the perimeter (total length of the walls) by 4 and round up to the nearest whole number. This is the number of panels you will need.

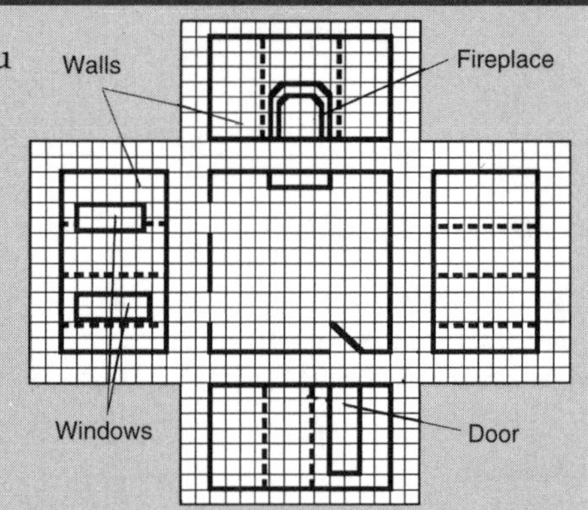

Conditioning the Panels

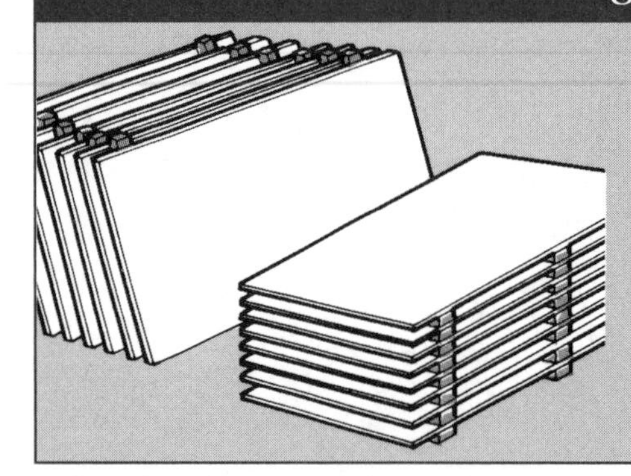

All paneling should be cured for a few days in the room where it will be installed, giving it time to expand or contract in response to the particular conditions of the location. Stack paneling flat with 1x2s between the sheets, or on the 8-foot edge, with separators between the sheets.

Arranging the Panels

Real wood veneers vary from sheet to sheet. If you are installing this type, stand the sheets up around the room and rearrange them into the most pleasing combination of grain patterns and hue. Putting dark panels with dark, and light with light, reinforces the simulation of plank paneling by disguising the sheets. Consider which walls in a room get the most and least light and put the lightest panels on the darkest wall.

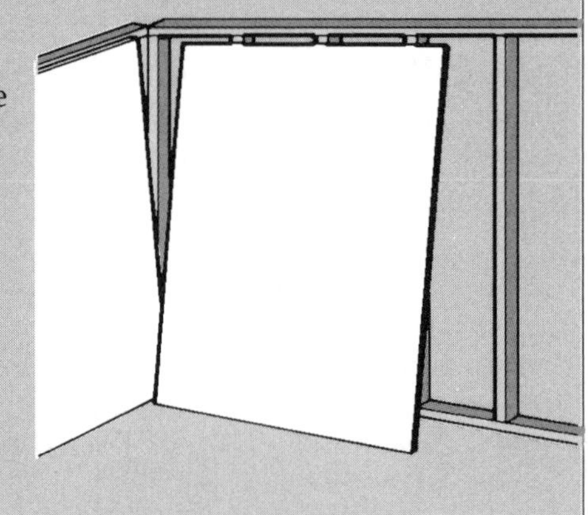

Cutting Paneling

Although paneling is easy to work with, it requires that you cut it carefully to avoid splintering the good face. The tool you use to make the cuts determines which side of the paneling you work on. Circular saws and saber saws are cut on the upstroke and should be used from the back of the panel. All saws that cut on the downstroke (handsaw, table saw, radial arm saw) should be used from the front. Use a plywood blade with six teeth per inch on a circular saw, a saber saw blade with ten teeth per inch, and a handsaw with a narrow set between cutting points.

Making Long Cuts in Paneling

You can be certain of straight cuts with a circular saw if you clamp a straightedge to the paneling as a guide for the saw. The edge of the temporary guide should be the same distance from the cutting line as the saw blade's distance from the edge of the saw's base plate.

Cutting Openings

Measure the position of an opening and transfer the location to the appropriate side of the panel. A quick way to do this is to rub the edge of the fixture that comes through the panel, with the panel exactly in place. Tap it over the spot with a hammer on cloth to protect the panel. Drill starter holes, large enough to start your saw, inside the corners of the outline.

Scribing

To fit a panel against an irregular surface, tack it to the wall a few inches away, checking that it is plumb, and use a carpenter's scribe to mark the edge of the panel. Cut the scribed outline with a jig saw or coping saw.

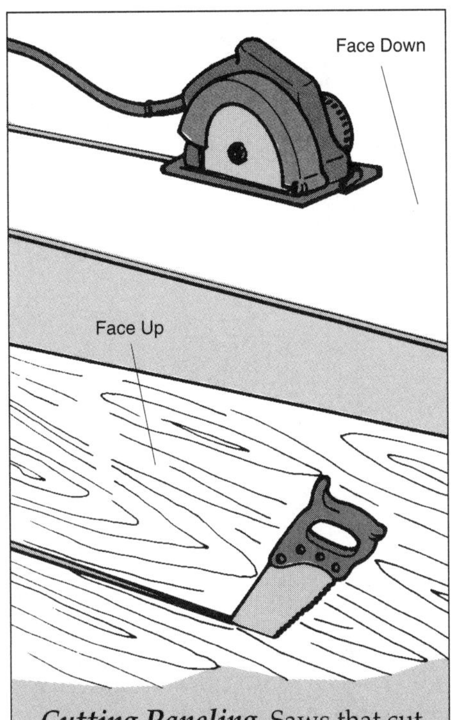

Cutting Paneling. Saws that cut on the upstroke are used from the back of a panel, those that cut on the downstroke are used from the front.

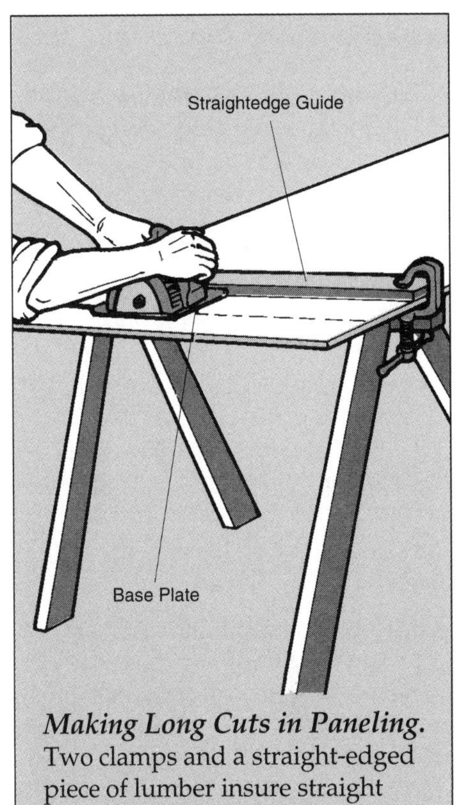

Making Long Cuts in Paneling. Two clamps and a straight-edged piece of lumber insure straight cuts. This jig is especially useful when cutting panels for a low wall.

Cutting Openings. Drill starter holes in each corner and use a keyhole saw or a saber saw to cut openings, using the starter holes to turn corners.

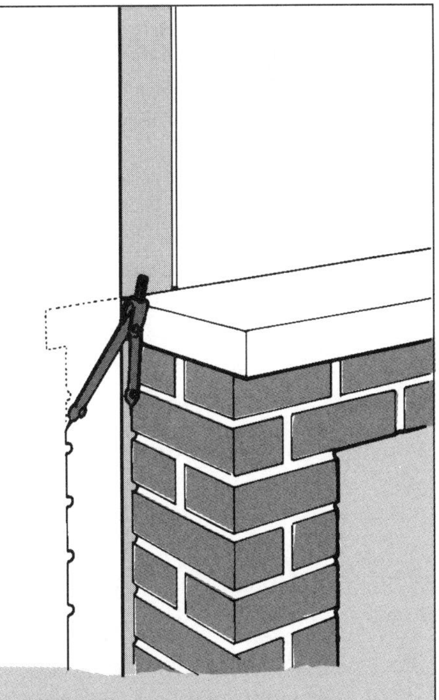

Scribing. The secret of cutting irregular shapes accurately is positioning the panel correctly before marking. Make sure it is secured.

Installing Sheet Paneling

1 Cutting the First Panel. Start at a corner. Check the ceiling and floor at the wall to see whether the room is square. If it is, measure the height of the wall and cut the panel 1/2 inch shorter. If it is not, find the distance in which the ceiling deviates and mark that on the appropriate side of the panel. Draw a cutting line to the opposite corner and make the cut. Measure the distance from the ceiling to the floor where both edges of the panel will sit, subtract 1/2 inch for clearance, mark the panel and cut. Use a plumb line to check the adjoining wall and cut to fit.

2 Applying Adhesive. Install paneling with either nails or adhesive. Use 3d nails (colored) when nailing into studs and 6d nails (colored) when nailing through wallboard into studs. Square the panel before nailing. For adhesive application, apply to furring. On a flat wall, cover surface with a random squiggle, and run a bead around the panel 1/2 inch from the edge.

3 Plumbing the Sheet. Set the panel against the wall, propped on wood shims and off the floor. Check the panel for plumb and correct the position. If the panel is not plumb, take it down and adjust it.

4 Curing the Adhesive. Many adhesives call for the panel to be held away from the wall to let the adhesive cure. Tack it in four places at the top and prop up at the bottom.

5 Concealing Seams. Use a felt-tip pen, the same color or darker than the grooves in the panels, to mark the surface along a seam before installing the next sheet. Install subsequent sheets as above.

6 Ending a Wall. When you meet the opposite corner of the wall, measure between the last panel and the corners at the top and bottom and transfer the measurements to the panel. Cut to fit, but not such a tight fit that you have to flex the panel.

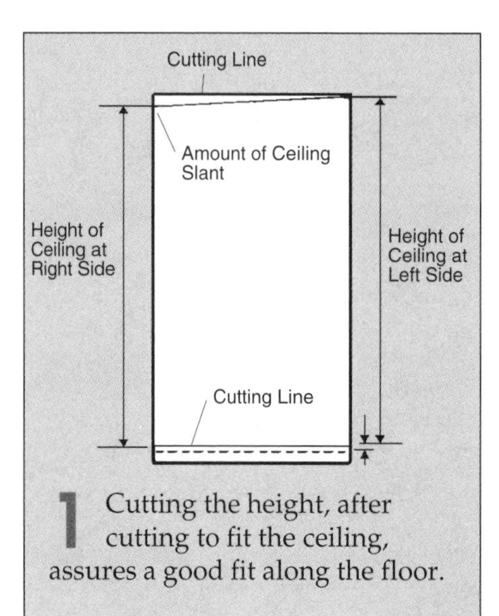

1 Cutting the height, after cutting to fit the ceiling, assures a good fit along the floor.

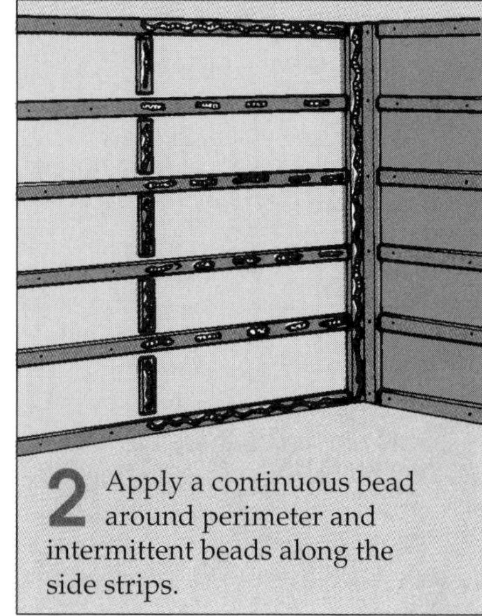

2 Apply a continuous bead around perimeter and intermittent beads along the side strips.

3 The first panel determines the position of all others, so make sure it is plumb.

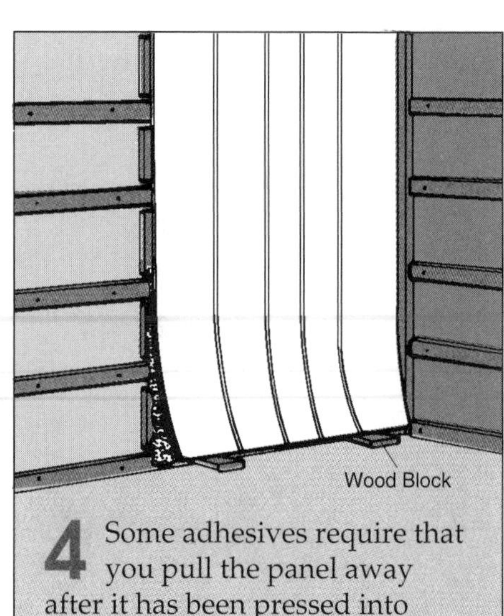

4 Some adhesives require that you pull the panel away after it has been pressed into the adhesive.

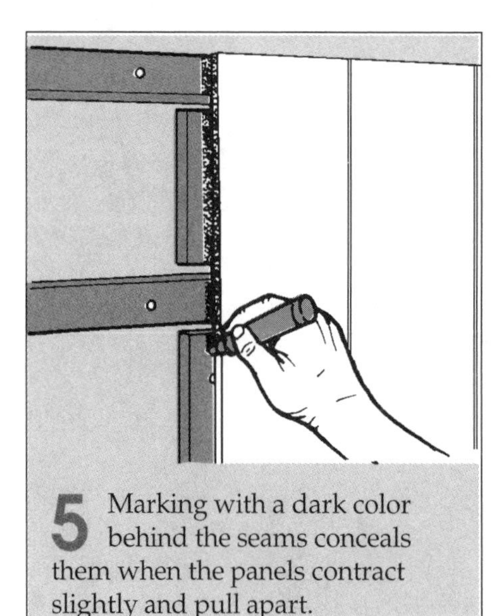

5 Marking with a dark color behind the seams conceals them when the panels contract slightly and pull apart.

6 The corner should fit closely, so measure at the top and bottom to cut the edge that meets the corner.

Creating a Herringbone Pattern

1 Cutting the Panels. You can achieve a dramatic effect by installing paneling in a herringbone pattern. Mark the sheets as shown. These cuts will produce a rearranged panel width of about 2 feet, 10 inches. This procedure will not work over a 16-inch-on-center frame. Measure the wall and plan the project so that you do not end with a noticeably narrow piece.

2 Fitting the Panels. Match up the pieces and attach. Repeat the procedure, matching the grooves of the next pair with the first.

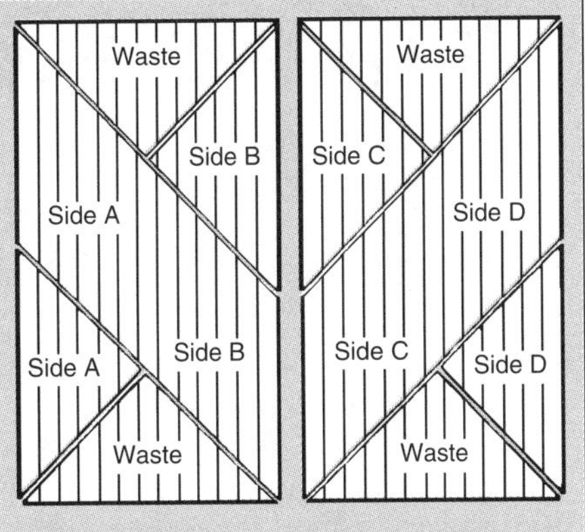

1 The letter code indicates the cuts for a herringbone pattern. Use a combination square to measure angles.

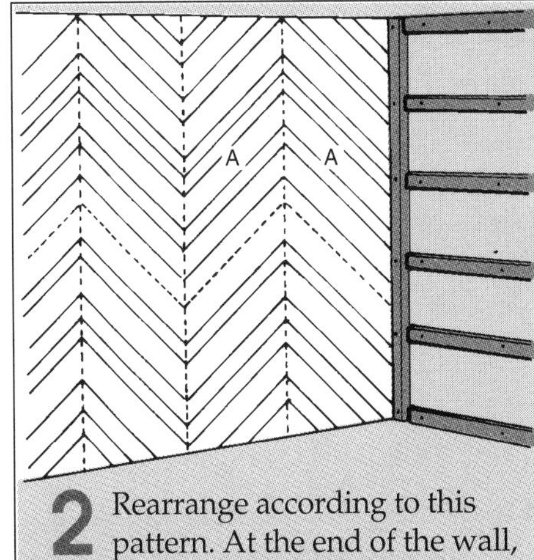

2 Rearrange according to this pattern. At the end of the wall, cut the panel to fit before cutting the diagonals.

Wainscoting

Paneling that covers only the lower 30 to 36 inches of a wall is called wainscoting and it is applied to the wall the same way as full sheets of paneling. Cut the paneling down to the desired height, measuring for each piece from a horizontal line on the wall, and install against the bare wall or over furring if necessary. If wainscoting is attached to a bare wall it can be capped with cap molding; if installed over furring, a combination of moldings may be required to fill the space between the paneling and the wall (see below).

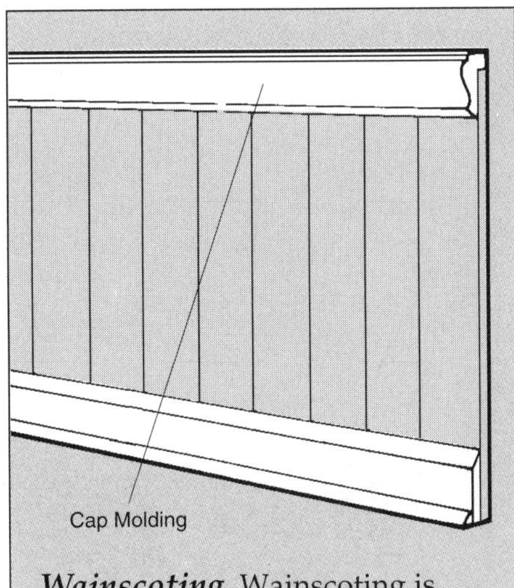

Wainscoting. Wainscoting is installed along a horizontal marked on the wall. Pieces should be cut to fit, measuring from the horizontal to the floor.

Covering High Walls

When panels must be stacked, either vertically or horizontally, to cover a wall, the seams created by butting the ends of panels can be concealed with molding or covered with strips of appropriately colored 1-inch board. The example on the right uses two pieces of paneling and molding to cover an extra-high wall.

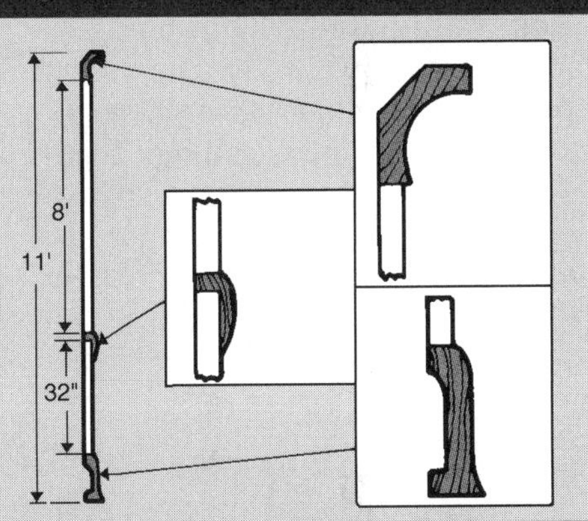

Fastening Trim for Hardboard Panels

Hardboard paneling can be attached with metal fastening trim available in four configurations—for meeting at inside corners, outside corners, ending at a corner, and meeting along a wall. Trim holds the panels in place and also provides a finished treatment. The trim can be cut with a hacksaw. Follow manufacturer's instructions—some pieces must be installed before a panel is seated, others are installed after a panel is seated. All trim must be installed plumb to yield a straight wall.

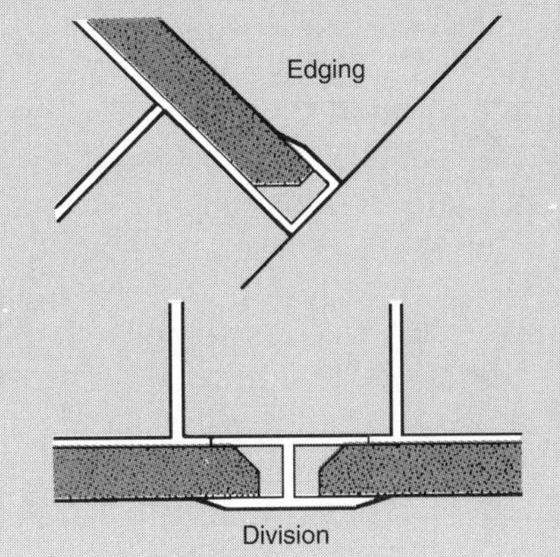

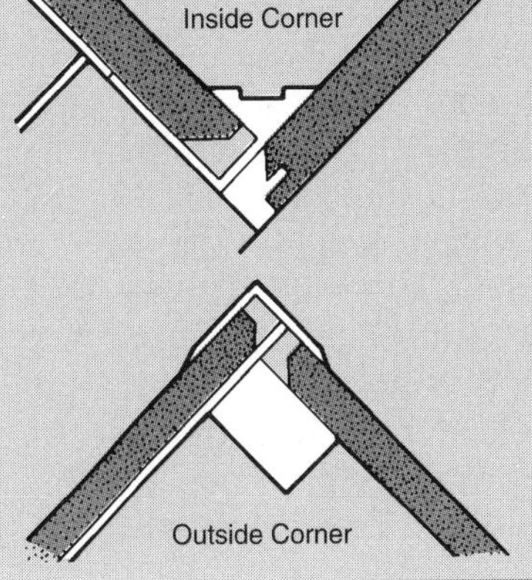

Working with Molding

Molding unites floors, ceilings, and walls and gives doorways and windows a decorative highlight. Molding also hides ragged junctions of a wall with the floor and the ceiling. Installing molding can be a simple job if you master the art of making various kinds of simple miter cuts to assure a tight fit.

For general use, you will find softwood molding in a great variety of shapes, as well as oak baseboard molding to match oak floors. To trim paneling, you can purchase molding with a finish that matches the paneling. Molding is designed for specific locations and uses. Baseboard molding and base shoe molding are used in conjunction to trim a wall at the floor; cove molding is used along the wall at the ceiling; casing is used around doors and windows; cap molding is used to cap wainscoting and corner molding for both inside and outside corners. It hides seams and softens corners.

Molding is available in standard lengths from 6 to 14 feet in 2-foot increments. Try to get lengths that will span walls from corner to corner, but if a wall is too long, you can splice pieces together. When measuring for a job remember to take into account the extra few inches you may need when mitering to meet an adjoining piece at a corner. To cut miters in molding you need a good miter box and a saw with enough teeth per inch to make fine cuts without splintering the molding.

Installing the molding is usually the last job in building or renovating a room. It is ordinarily painted, or finished in some other way, with a different color than the walls. This should be done and the walls should be painted before the molding is attached. A common exception to this rule occurs when baseboard molding is painted the same color as the room and the base shoe is stained a different color. Then the baseboard is painted with the wall.

Types of Molding

Molding is milled in dozens of different shapes. Casing (left) for doors and windows, baseboard (center) for floors, and cove (right) for ceilings are the most familiar moldings. Because techniques for milling molding are not perfectly standardized, it is better to buy all pieces from the same milling lot if you can, to avoid fractional differences in size.

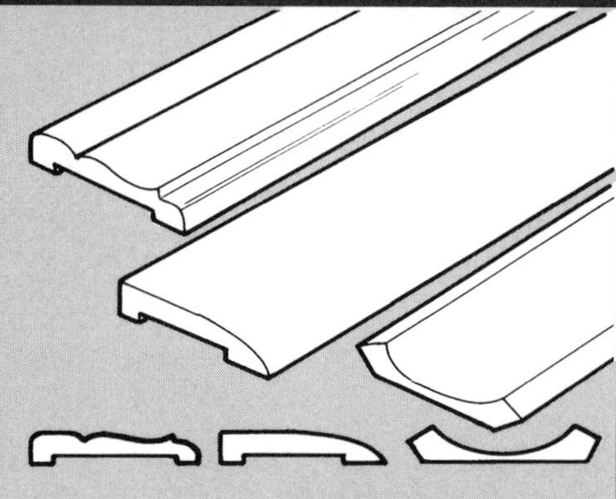

Cutting Miters

1 Cutting Baseboard. Using a miter box is quite simple, but you must take care to make the cuts in the right direction or you may be left with a piece too short to recut. For baseboard molding, mark the point at which it is to be cut at the top and put the molding in the miter box right side up with its back against the back of the box. When cutting an inside corner, set the saw guide so that the back of the molding (the wall side) will be longer than the front side. When cutting an outside corner, the saw should be set so that the front side will be longer than the back. Double-check the orientation of the molding in the miter box and the position of the saw before cutting. To protect the face of the molding from splintering, you can apply a piece of masking tape down the cut. Test fit the molding before attaching it. You can adjust the angle by shaving down any face that causes a gap in the joint—do this cautiously because it shortens the molding.

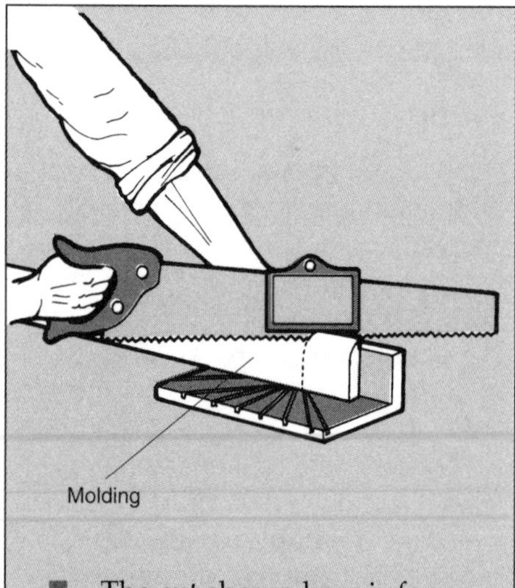

1 The cut shown here is for an inside corner; the front face will be shorter than the back, or wall, side.

2 Cutting Cove Molding. The procedure for cutting cove molding (for ceilings) is essentially the same for cutting baseboard, except that the molding should be set in the box upside down so it can be seated firmly against the bottom and the back. As with cutting baseboard, the wall side is the long side for inside corners; the short side for outside corners.

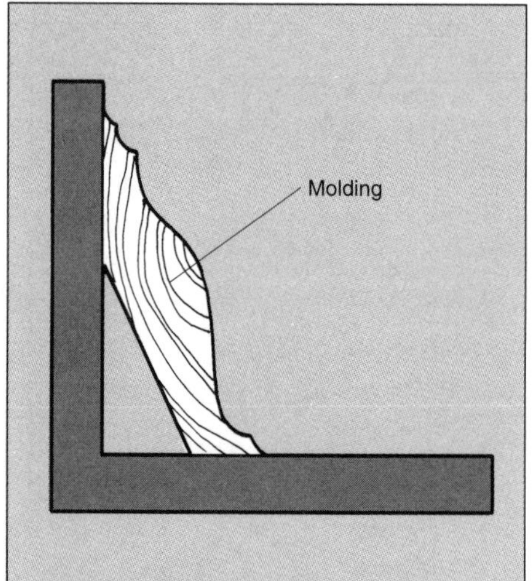

2 Mark the bottom edge of cove molding and set it in the miter box upside down to make the cut.

Cutting Molding to Butt at Right Angles

1 Cutting the Molding. When butting molding with a figured face, at a right angle to other molding already in place, you must shape one piece to fit the contours of the other by making two cuts. First measure, mark, and cut the piece as if you were mitering it to fit into the corner.

2 The Fit. Mark a second cutting line along the edge of the cut on the face and cut this, following the curves, with a coping saw held vertically. The second cut shapes the piece to fit over the face of the molding it abuts.

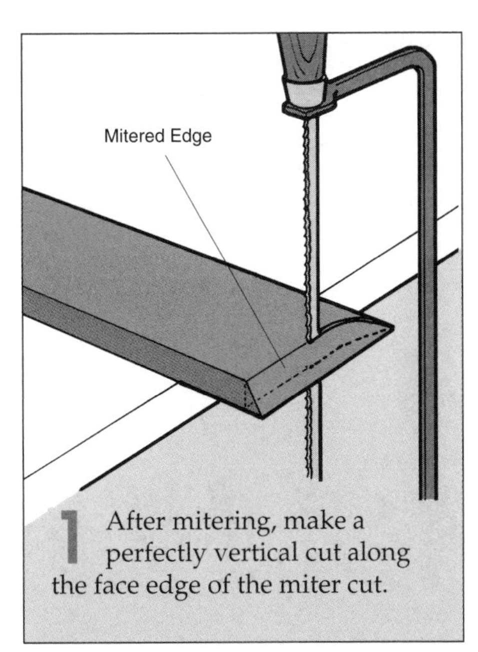

1 After mitering, make a perfectly vertical cut along the face edge of the miter cut.

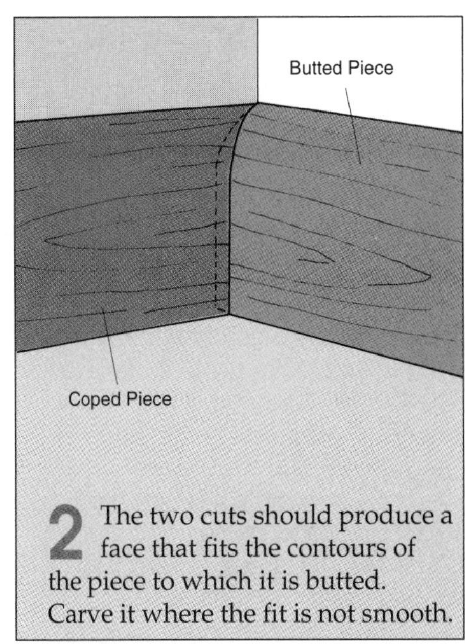

2 The two cuts should produce a face that fits the contours of the piece to which it is butted. Carve it where the fit is not smooth.

Nailing Molding

Depending on the size of the molding, use 6d or 8d finishing nails to attach baseboard, ceiling molding and casing; use 4d finishing nails for base shoe molding and drive them into the flooring. Nail into studs and top and bottom plates to assure that the molding will not pull away from the wall. Sink nail heads with a nail set and cover. Tighten corners with two 4d nails into the wall on either side; one nail for base shoe molding.

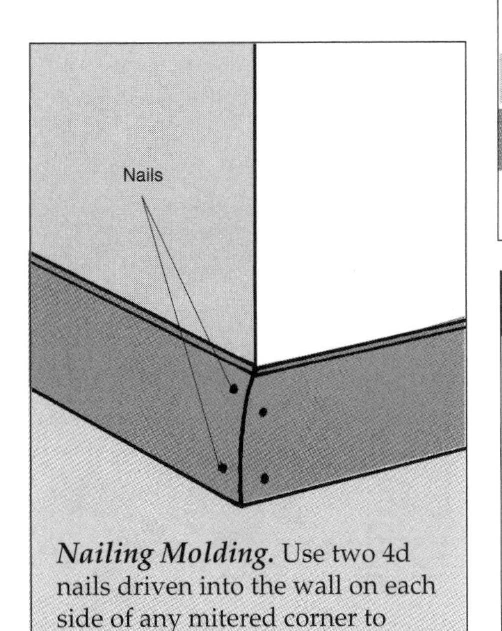

Nailing Molding. Use two 4d nails driven into the wall on each side of any mitered corner to keep the corner from spreading as the house settles.

Lapped Joint for Molding

If you must join pieces of molding along a wall, miter both pieces to make a smooth lap joint that can be tied together with a single nail.

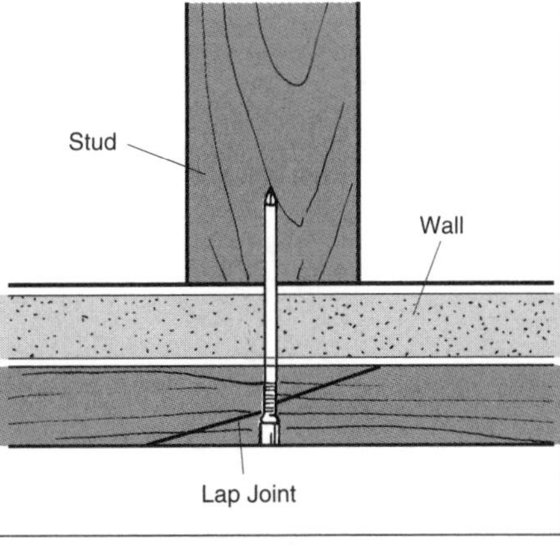

Base Molding at Doors

Where a baseboard butts against a door jamb but the base shoe protrudes, mark the shoe before installation and bevel the end to smooth junction with door jamb.

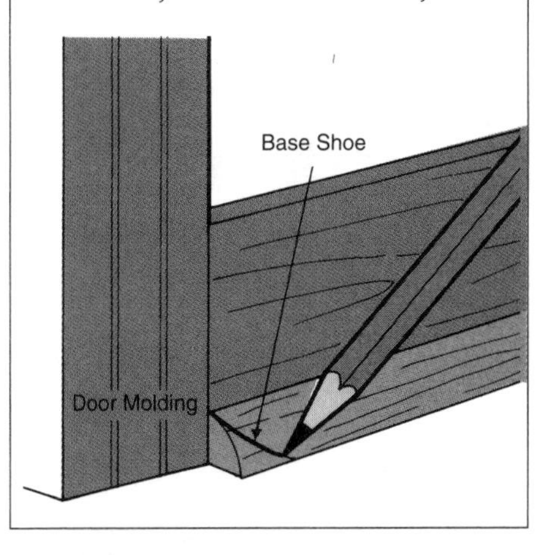

Trimming a Window

Windows without sills can be framed like pictures with four pieces of casing, mitered at all corners. Windows with sills are trimmed on the sides and top with casing and on the bottom with a piece of window stool (the technical name for the sill) and an apron. The stool should be installed first. Window jambs must be built out before molding can be attached, when a new surface has been put on a wall.

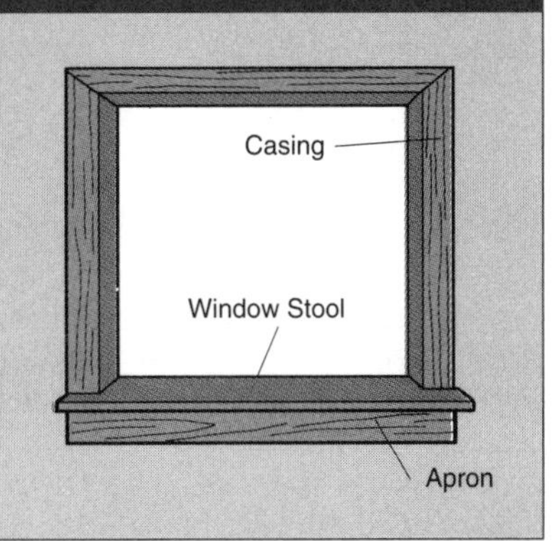

Ceramic Tile Walls

Ceramic tile is an attractive wall surface anywhere and a practical one in bathrooms and kitchens because it is waterproof, fireproof, durable, and easy to clean. The variety of tiles available make it a decorating favorite.

Tile is clay that has been fired; it is produced in a variety of surfaces. The most important distinction between types of tile is whether it is glazed or unglazed. Glazed tile, available in a matte or shiny finish, is impervious to stains but it can be scratched. It is the standard tile around sinks and tubs. Unglazed tile, made only in matte finish, picks up stains from grease and oil but resists scratching. Both kinds can be used on walls. Flat tiles are called field tiles and those shaped to fit around corners and edges are called trim tiles. Tiles larger than a few square inches are sold loose; smaller tiles can be purchased in sheet form with a few square feet of tile bonded to a thin webbing on the back.

To install ceramic tile, you need a clean surface that is sound and flat. Wallpaper and other flexible coverings should be stripped before tiling an old wall. Unsound areas should be repaired. Both water-resistant wallboard and plywood make good surfaces for tiling in high moisture areas. The wall surface is marked with guidelines to position the tile, then covered with adhesive into which the tiles are pressed. The joints between tiles are then sealed with grout.

Tiling requires only a few special tools: a tile cutter for cutting straight lines in tile and tile nippers to cut odd shapes (both can usually be rented); a notched trowel for spreading adhesive (check with your dealer to match the trowel with the job); and a rubber float for applying grout. You also will need measuring tools and a level.

Patterns for Laying Tile

There are two basic ways to lay tile on a wall: Build the tiles up from the center of the wall in a pyramid shape (left), or start by laying the length of the bottom row, then work from one corner at the bottom diagonally up the wall (right).

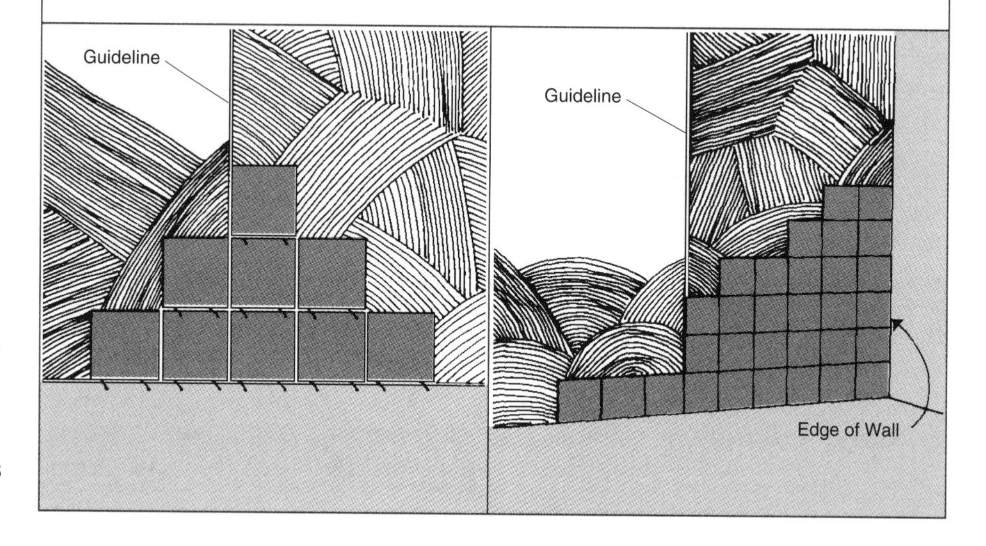

Planning the Job

- If you are tiling to the top of a wall, and the ceiling is level, measure the height of the wall at both corners;
- If you are ending the tile part way up a wall, mark a horizontal line across the wall at the desired height and measure from the corners to that line;
- If you are tiling down to a floor and ending the tile with trim pieces, you should mark the horizontal guideline for tile position at the height of a trim tile plus a field tile, above the floor;
- If you are tiling from the edge of a tub, establish the horizontal guideline at the height of a single tile above the tub;
- If you are tiling around three sides of a tub, measure up the wall at each corner to the height you want tiled (a minimum of one tile above the shower head), then extend horizontal lines from those points to the center of each wall and note the distance of any gap where the lines meet. This will tell you how far out of level your tub is (see below).

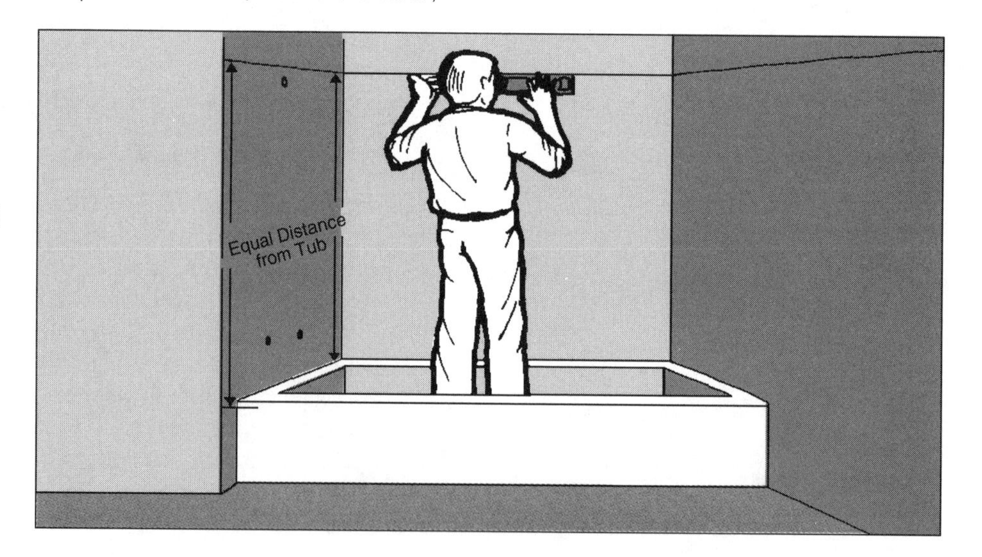

Marking Guidelines on Wall

1 Locating a Vertical Guideline. Tiling from the middle of a wall is the easiest method. Your first measurement should be to determine the center of the wall. Lay out a run of tiles, including the additional spacing for grout for tiles with straight edges, but no additional space for tiles with lugs on the edges that set the spacing when butted. Mark directly on the wall. Check whether the last tile on either side of the midpoint is more or less than half a tile wide. If it is less, mark half the width of a tile to the right or left of the midpoint and use this as the position of the vertical guideline.

2 Marking the Vertical Guideline. Use a carpenter's level to mark the vertical guideline at the proper point.

3 Establishing a Horizontal Guideline. Check whether the floor or tub is level. If it is less than 1/8 inch off from one end to the other, mark the horizontal guideline from the highest point. If it is more than 1/8 inch off, mark the horizontal from the low end. If you are meeting a floor or counter with trim tile, put a piece of trim in position and a field tile on top, held at the proper grout spacing or butted on lugs, and then put a level on the lugs or equal to the grout space above. Mark along the bottom of the level and extend the line to the length of the wall. If you are meeting a tub with field tile, measure from the top of a single piece of tile set on the lowest point. If you need to fit tiles against a ceiling, measure down from the top and use the method above to establish the height to which the first row of tiles must be trimmed. Trim a few tiles to that height, and use one to establish the horizontal above the tub rim.

4 Marking for Accessories. If you are mounting the kind of soap dish that sits flush to the wall, or other similar ceramic accessory, find the position where the piece is to be located and mark off the dimension of the part that will sit against the wall. Most pieces have flanges that fit over surrounding tile.

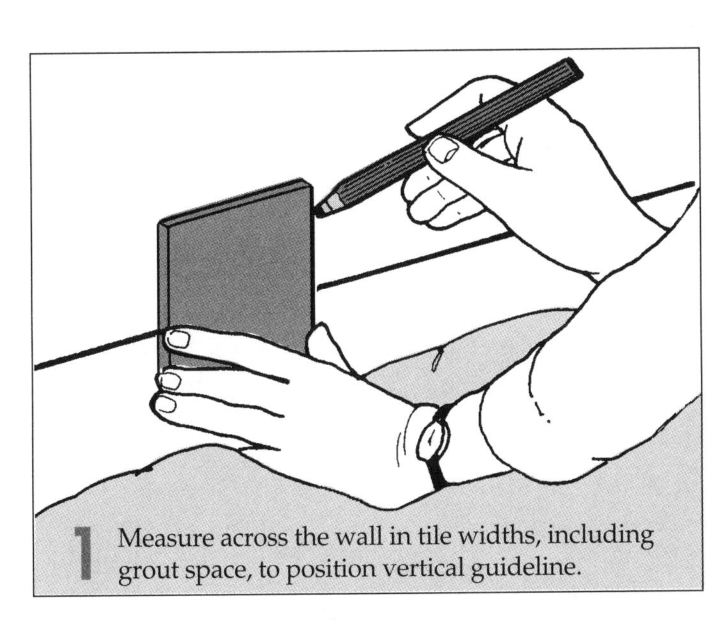

1 Measure across the wall in tile widths, including grout space, to position vertical guideline.

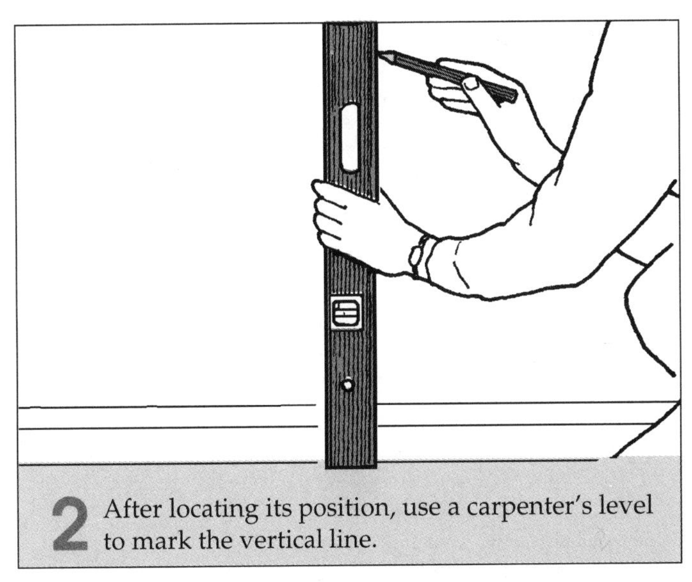

2 After locating its position, use a carpenter's level to mark the vertical line.

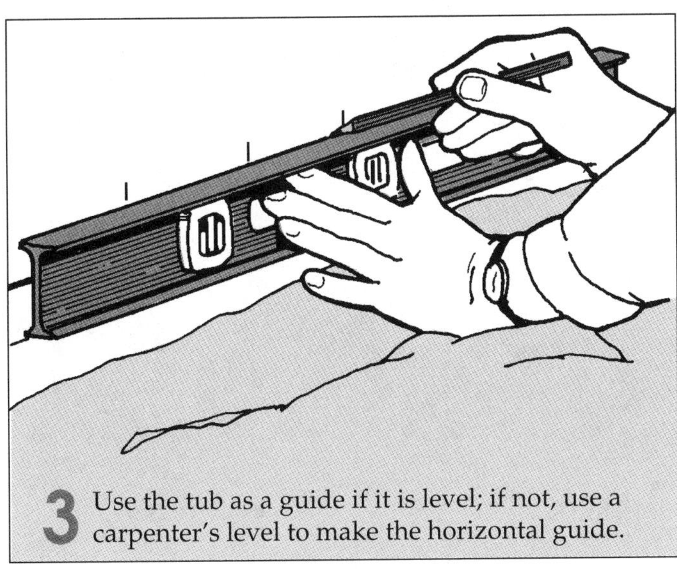

3 Use the tub as a guide if it is level; if not, use a carpenter's level to make the horizontal guide.

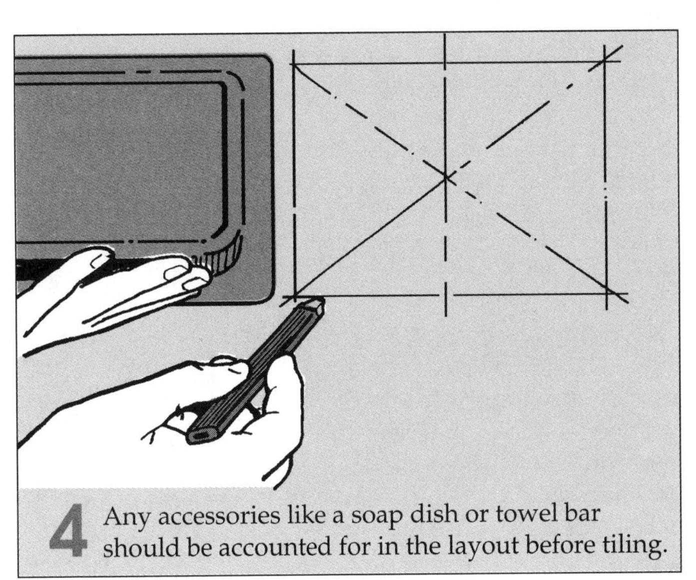

4 Any accessories like a soap dish or towel bar should be accounted for in the layout before tiling.

Installing Ceramic Tile

1 Applying Adhesive. Tile can be bonded to the wall with a mix-it-yourself cement-based adhesive or, with greater ease, a premixed mastic adhesive. The basic method of applying a mastic adhesive is to scrape it across a wall off the long edge of a notched trowel. The depths of the bottoms of the valleys at or near the wall surface should be consistent. Start at the guidelines and work your way out, leaving the lines visible.

2 Finishing Adhesive. If you are working on a small wall, cover the entire surface with adhesive; if you are tiling a large surface, work in smaller areas. Cover the surface entirely and evenly. The adhesive dries slowly, allowing time to make adjustments. Do not cover such a large area that the adhesive loses its tackiness before you get to it.

3 Laying First Tiles. Line up the first tile along the vertical guideline and tip it into position. Give the tile a slight twist to spread the adhesive beneath it more evenly. Settle the tile half a grouting space away from the horizontal and vertical lines if it has no lugs; with the lugs on the lines if it does. Set a second tile on the other side of the vertical line in the same manner. If the tile has lugs, butt them. If there are no lugs, position for the grouting space by hand, or use an object of uniform thickness such as a piece of cardboard, a sliver of wood or a nail. The only rule is that the spacer must be the thickness of the gap you want. Tiles without lugs should be supported with nails under each one. Tiles with lugs require nails only under the bottom row to keep them in position.

4 Completing the Wall. Use either the pyramid or diagonal method (see page 40), to position subsequent tiles. If the tiles are laid carefully along the guidelines, their straight edges will serve as in-place guides for the next tiles, and they in turn for subsequent ones.

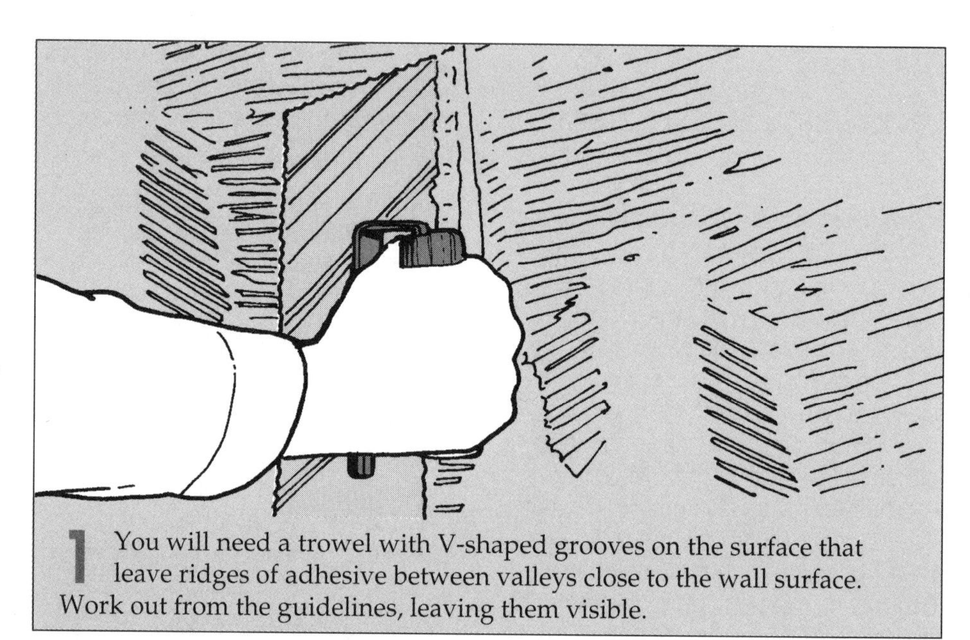

1 You will need a trowel with V-shaped grooves on the surface that leave ridges of adhesive between valleys close to the wall surface. Work out from the guidelines, leaving them visible.

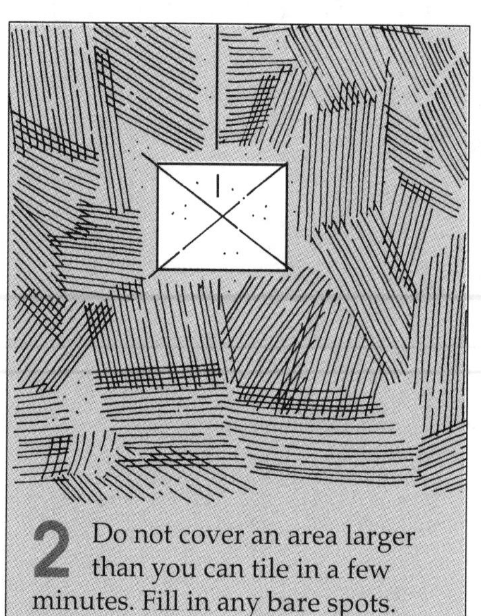

2 Do not cover an area larger than you can tile in a few minutes. Fill in any bare spots.

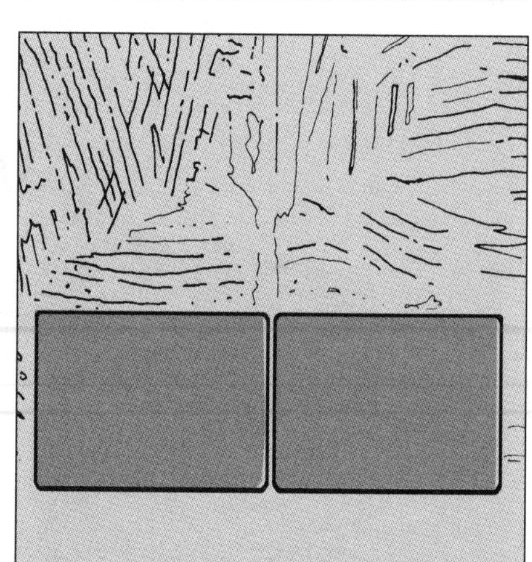

3 Place tile at intersection of vertical and horizontal guidelines. Seat its base and tip tile into position.

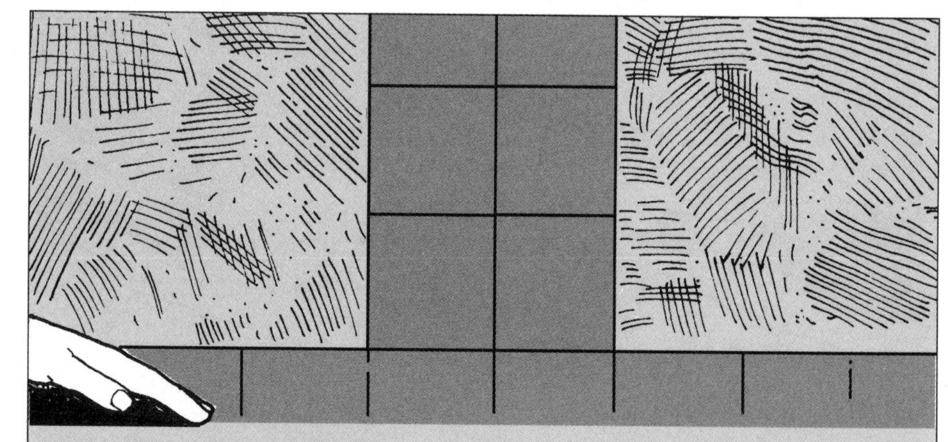

4 Using the first tile and the guidelines for reference, seat the second tile and continue in whatever pattern you choose. Constantly check positions of tiles to be sure they are aligned.

Mounting a Soap Dish

A soap dish or other ceramic hardware should be the last piece installed. The surface should be free of tile adhesive. Soap dishes must be fixed to the wall with care because they may be subject to unusual stress and should not give way suddenly. Use epoxy putties or special mastic-like adhesives designed to do the job.

Mounting a Soap Dish. A soap dish, or any other fixture that may have to bear weight, is installed with a different adhesive than wall tile. The adhesive is applied to the soap dish, then the dish is stuck onto the wall. Follow manufacturer's instructions on drying which may take several days.

Cutting Ceramic Tile

1. Use a tile cutter to make straight cuts. The cutter holds the tile in place while you score the surface with a wheel at the end of a handle mounted on a fixed track. Run the cutting wheel back and forth, applying some pressure to cut the surface, then tip the handle back to break the tile along the score line. Always cut tiles with ridges on the back in the direction the ridges run.

2. The cut is likely to have some rough edges which should be squared off with a piece of metal plaster lath.

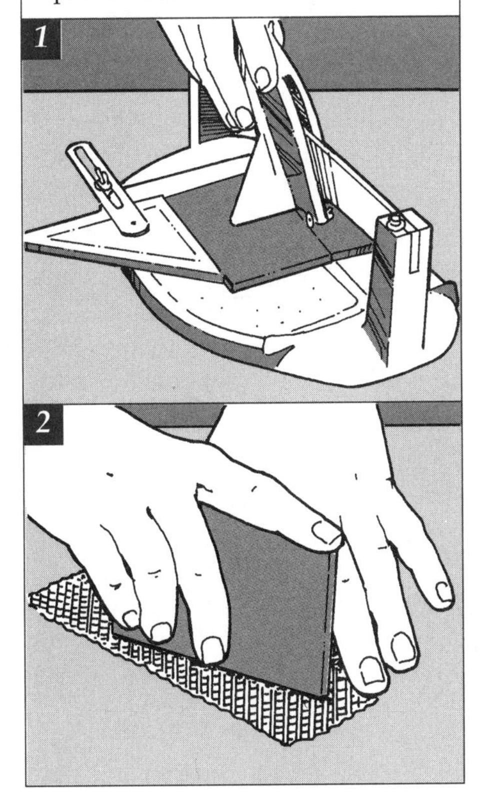

Cutting Tiles to Fit Odd Shapes

1 **Starting a Cut.** To fit tile around such edges as a shower head or a pipe, you must cut it bit by bit with tile nippers. This takes a strong wrist and some patience because several nips through the very tough tile may be needed to carve out the shape.

2 **Finishing a Cut.** Most pipe areas and holes you must cut around will later be covered by an escutcheon or flange concealing the edges of the cut, so you need not worry about a ragged edge as long as the cut is large enough. To fit around pipes, cut the tile into the two sections and nip out a semicircle, the size of the pipe, on both meeting edges. If you need a very precise circular cut in a tile, you can drill it with a carbide-tipped hole saw, cutting from the back of the tile.

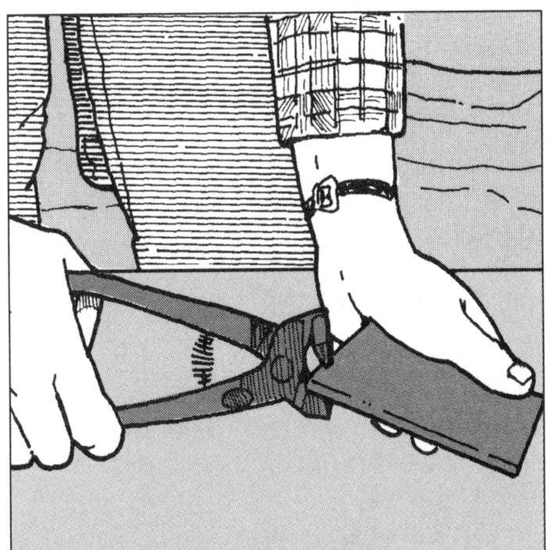

1 A title cutter will only cut straight lines; for odd cuts you must use tile nippers.

2 Nip away pieces of tile to the cut line. This need not be perfectly smooth.

Mirror Tile Walls

Mirror tiles offer a quick and relatively inexpensive way to change the character of a room. A mirrored wall will greatly increase the apparent size of a room while increasing its brightness with reflected light. There are several varieties of mirror tile to be found. The three most common are lightweight 1-foot-square tiles that are attached to the wall with double-sided tape; heavier tiles of various sizes, usually with beveled edges, that are attached to a wall with special mastic and with hardware to hold them in place; and longer strips of mirrored glass attached with clips.

The two main challenges in installing mirror tiles is to get them firmly fixed to the wall so they will not shift and crack or fall off, and to get them installed evenly so that reflection will not be distorted. Install mirror tile only on walls that are flat and even. Lightweight tiles that can be attached with tape are thin enough to cut at home, but heavier tiles and strips are too thick to cut yourself. With the thicker tiles, plan to leave an untiled border. Heavier tiles must be glued with a mastic made especially for mirrors so that the silvering on the back is not damaged. Check with your lumber or hardware dealer for a selection of adhesives for this job.

Installing Heavier Tile

Paneling a wall with heavy mirror tile requires either clips or special screws, and mastic to hold the sections to the wall. Different products require different hardware so plan the job with your dealer.

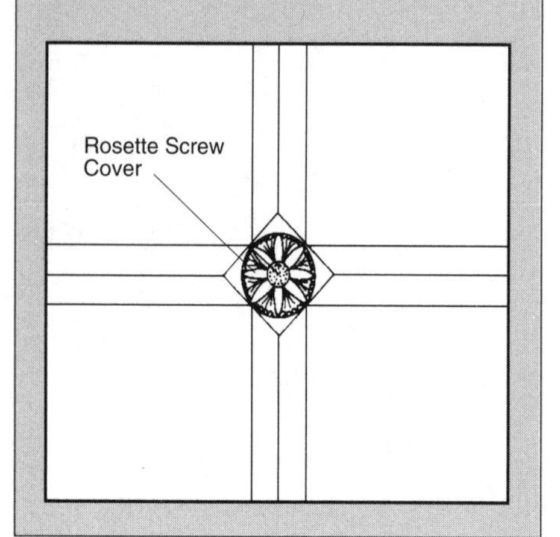

Installing Lightweight Tile

1 Marking Reference Lines. Measure the length of the wall and mark the midpoint at about eye level. If the wall is an even number of feet long (or closer to even than to odd), use a carpenter's level and straightedge to draw a horizontal line, and a plumb bob and chalkline to snap a vertical line, both through the midpoint. If the wall is an odd number of feet long, draw these lines to intersect 1 foot to the right or left of the midpoint. Be sure that the lines are true horizontal and vertical—they are the reference from which all tiles will be installed.

2 Applying Tape. Peel back the protective cover from one side of squares of double-sided tape and stick one in each corner of the mirror. Be careful not to touch the exposed adhesive because oils from your fingers can impair its bonding ability. Be sure that the wall surface to which the tiles will be stuck is completely clean and dust-free. Remove the other protective cover from the tape squares and stick the first tile onto the wall at one of the four corners of the intersecting reference lines. Install additional tiles, quadrant by quadrant, taking care that they are butted perfectly.

3 Cutting the Tile. When you must cut a tile to fit along the edge of a wall or around an obstruction like a light switch, mark the cut on the mirror's surface, then score the mark with a glass cutter pulled along a straightedge and snap the cut.

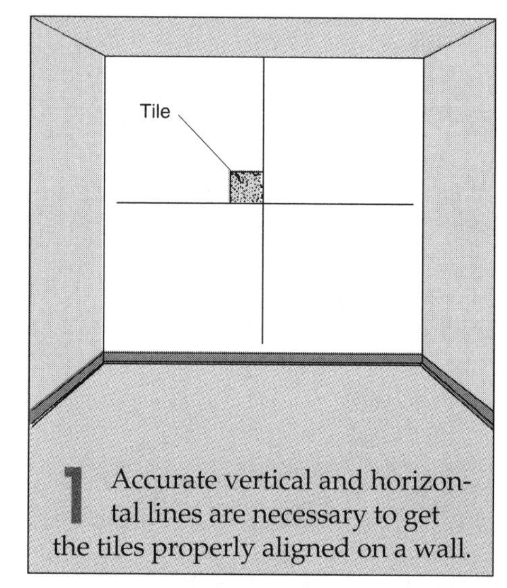

1 Accurate vertical and horizontal lines are necessary to get the tiles properly aligned on a wall.

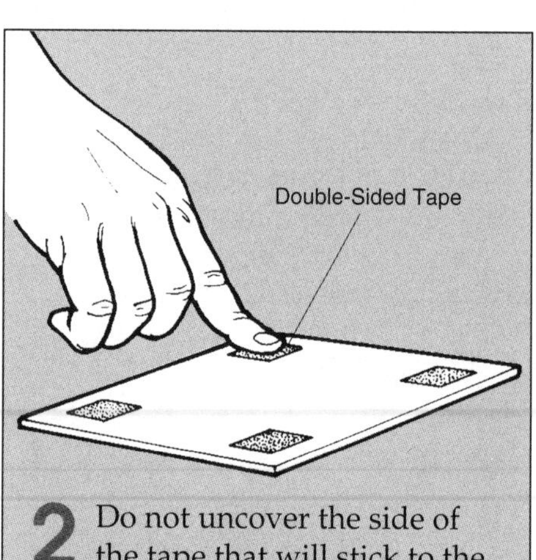

2 Do not uncover the side of the tape that will stick to the wall until you are ready to position the tile.

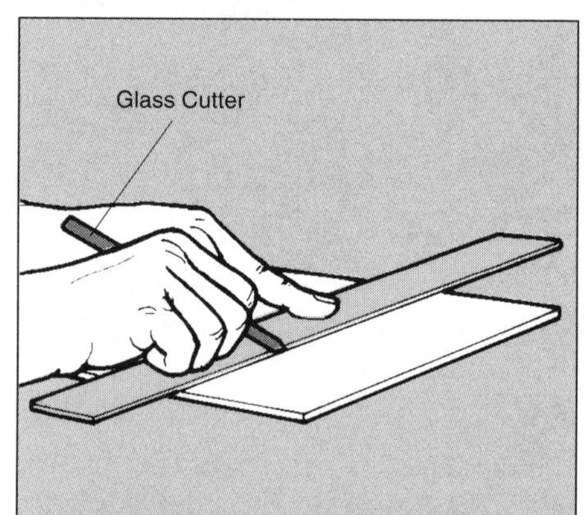

3 Score the tile with a glass cutter, using steady pressure in one swift motion. The glass will break along that line.

Painting Walls

The most important factor in any painting job is how well you prepare the surface. For paint to adhere properly, it must be applied to a clean, dry surface—free of dirt, dust, grease and flaking paint. Also, paint does not fill in defects. Cracks, dents and popped nails will show as clearly after painting as before. The care you invest in cleaning, scraping and patching will make your painting worthwhile.

The most common paints are latex and alkyd. Latex paints are water-based, odorless, quick to dry, easy to clean up, and relatively inexpensive. Alkyd paints, often inaccurately referred to as "oil-based," are synthetic-based, adhere to a variety of surfaces better than latex and, provide a richer, more durable finish. They are slow drying, require chemical solvents for thinning paint and cleaning tools, require that one works in a ventilated area, and are more expense. In addition to the basic tools discussed below, you will need a painting guide or edger, paddles for stirring paint, drop cloths, a step ladder and a supply of rags.

Brushes

Brushes are made with either natural or synthetic bristles. Natural bristles should never be used with a water-based paint; the bristles absorb water and become clogged. When shopping for brushes of any size, check the following points: The handle should be comfortable and easy to clean; the brush should feel full when gripped around the bristles; the bristles should be flagged (split), should fan out slightly, should not clump when pressed against your palm, and should spring back into position afterward. Tug the bristles; they should not come loose. Most interior painting jobs can be handled with three sizes: 1- or 1½-inch-trim brush, 2- or 2½-inch sash brush, and 3- or 4-inch brush for large areas. Do not buy larger brushes to make the job go faster. You will find them unwieldy and too heavy after a few minutes.

Rollers

A roller consists of two standard-size parts: the frame and the cover. Covers vary as to thickness and composition of the nap (most are nylon). Short nap, 1/4 inch thick, applies a thin, smooth layer of paint suitable for glossy paints. Medium nap, 3/4 inch thick, holds more paint and applies it with a slight stipple, making it the choice for most interior work. Long nap, 1 inch thick, deposits a large amount of paint and is designed for porous surfaces. Most roller frame handles are threaded inside to accept an extension handle. You will need a roller pan for loading the roller with paint. Also available are electric paint feeders that pump paint through a long hose to a specially designed roller.

Pads

Painting pads are easy to use, fast and versatile. The pads can paint corners, cut-in and trim, as well as paint large areas. Some pads are designed with guide wheels to aid in trimming and may be used with extension poles. All require a tray for loading paint.

Sprayers

For big paint jobs there are both compressed air and air-less sprayers. An air compressor unit generates air to spray paint. The air-less sprayer pumps the paint, at great pressure, through the nozzle. Both methods have considerable overspray and require careful masking, ventilation, a respirator and safety glasses.

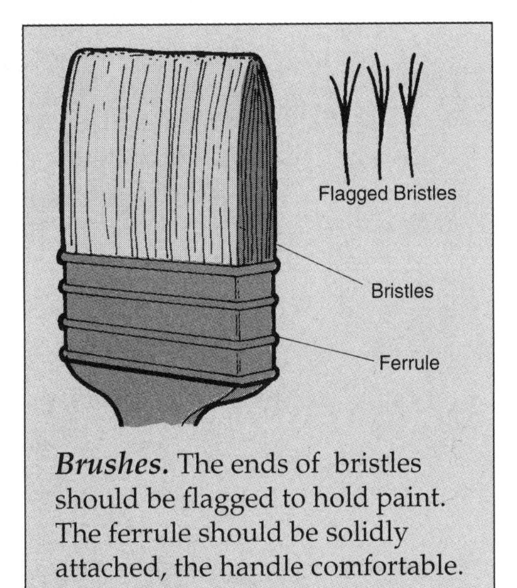

Brushes. The ends of bristles should be flagged to hold paint. The ferrule should be solidly attached, the handle comfortable.

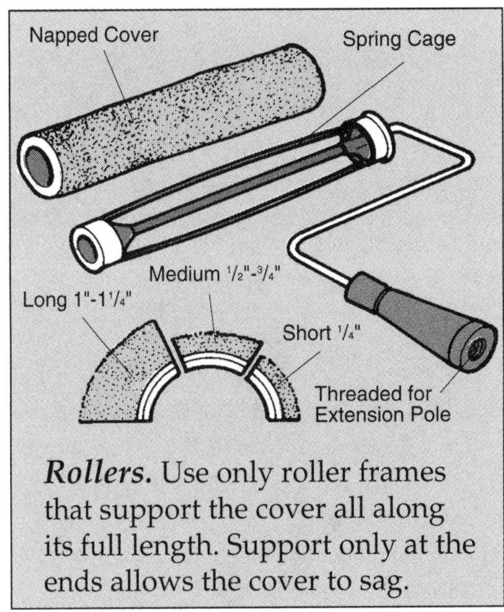

Rollers. Use only roller frames that support the cover all along its full length. Support only at the ends allows the cover to sag.

Pads. Paint pads may be used for flat surfaces as well as trim. Extension poles attach to threaded sockets.

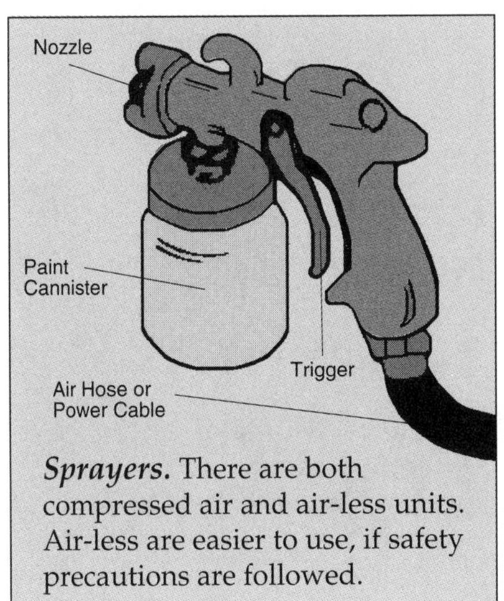

Sprayers. There are both compressed air and air-less units. Air-less are easier to use, if safety precautions are followed.

Scraping & Stripping Paint

Old paint that is peeling, blistered, or "alligatored," must be scraped or stripped entirely. Loose paint should be scraped with a paint scraper—if the flakes leave depressions, fill with joint compound and smooth them. Badly deteriorated paint, common on woodwork, can be stripped by melting it with a heat gun or by applying a chemical paint remover.

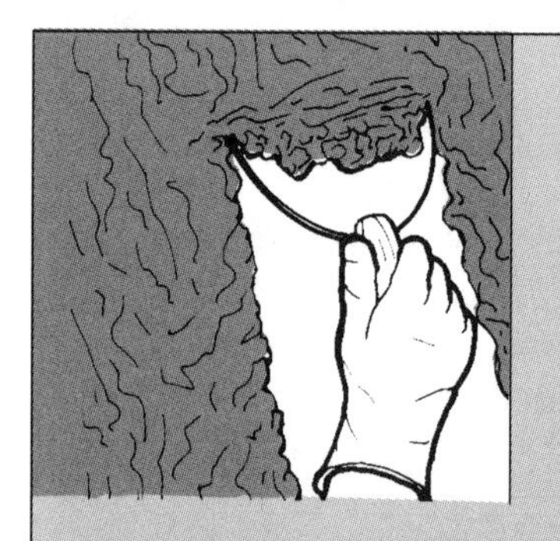

Scraping & Stripping Paint. Chemical paint removers soften deteriorated paint so that it can be easily removed. Observe manufacturer's cautions.

Stripping Wallpaper

Wallpaper can be painted over if there is only one layer on the wall, it is in good condition, and is not vinyl. If there are several layers on the wall, or if it is bubbled, peeling, or loose, it must be removed. You can do this in one of two ways: with a rented wallpaper steamer or with chemical agents.

Stripping Wallpaper. Wallpaper is first loosened by wetting, then peeled from the wall with a broad scraper. A wallpaper steamer will speed the job.

Getting Ready to Paint

A room ready to paint will look something like the one shown here—use the picture as a visual checklist of basic preparations. The walls have been cleaned, slick spots roughened with sandpaper and old paint has been scraped and patched. The floor and any contents of the room that cannot be removed are completely covered. Switch and outlet face plates are removed from the wall, but are left in the room. Wall and ceiling light fixtures are either removed or loosened from wall or ceiling to be enclosed in plastic bags. Other hardware such as door knobs, picture hooks, and thermostat covers have been removed.

The order in which you should paint an entire room is as follows: 1) ceiling, 2) walls, 3) trim, 4) doors, and 5) windows.

If you are painting a surface that has never been painted before, you must first prime it. There are both latex and alkyd primers, but new wallboard (and wallboard patches) must be primed with latex.

Handling Paint

Preparing the Can. Use a hammer and nail to tap holes through the rim of the paint can to prevent paint from accumulating.

Loading Paint. Don't dip the brush more than half way up the bristles. Tap it lightly on the rim to shed the excess.

Using a Roller

1 Spreading the Paint. Roll the roller through bottom of paint pan, where the paint should not be more than 1/2 inch thick, distributing paint evenly over entire roller. Start laying paint on the wall in a zigzag. Go back over the zigzag with parallel strokes at a 90-degree angle.

2 Finishing the Paint. Without reloading the roller, finish this area by carefully rolling up and down, overlapping the strokes slightly, or side to side, if you have decided to do your finish strokes that way. Start and stop your strokes gently so as not to leave roller marks on the wall.

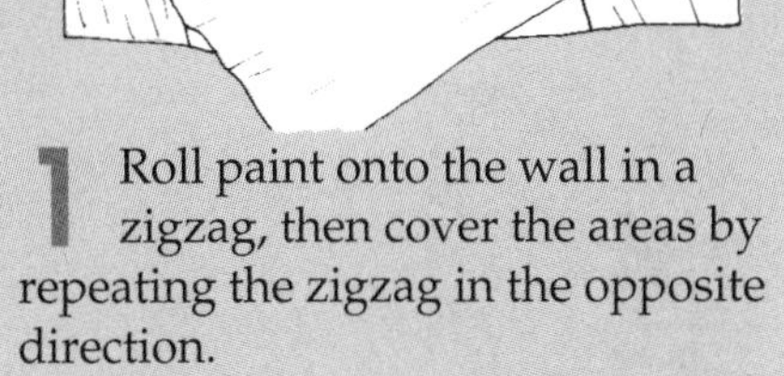

1 Roll paint onto the wall in a zigzag, then cover the areas by repeating the zigzag in the opposite direction.

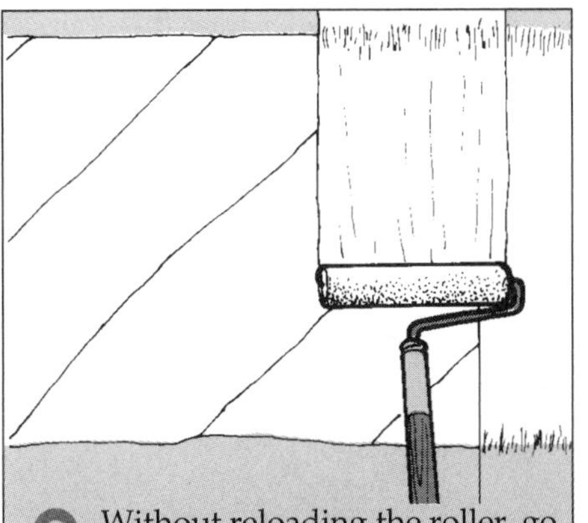

2 Without reloading the roller, go over the painted area with parallel up and down strokes to smooth the paint.

Cutting In

Making a clean edge where a ceiling meets a wall of another color or along an edge where paint stops is called "cutting in." Use a trimming brush, start the paint below the line, and pull the brush just along the line. When painting a wall and ceiling different colors, the standard approach is to let the lighter of the two colors overlap this edge, cutting in the darker of the colors over it.

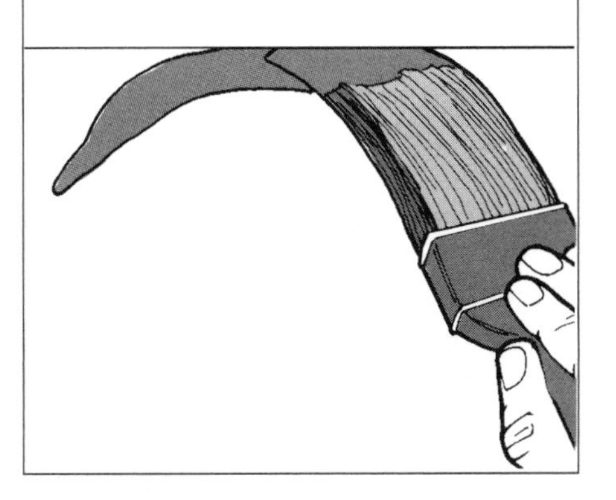

Painting Double Hung Windows

1. You can mask glass in windows with tape laid up to the edge of the strips that divide the panes or you can paint the strips freehand. The method for painting a double hung window is to lower the top sash and raise the bottom sash. Paint the outside sash first as far as you can reach, then paint the inner sash.

2. Reverse the position of the windows, but do not close either sash all the way. After the windows are painted, paint the sill and casing around them.

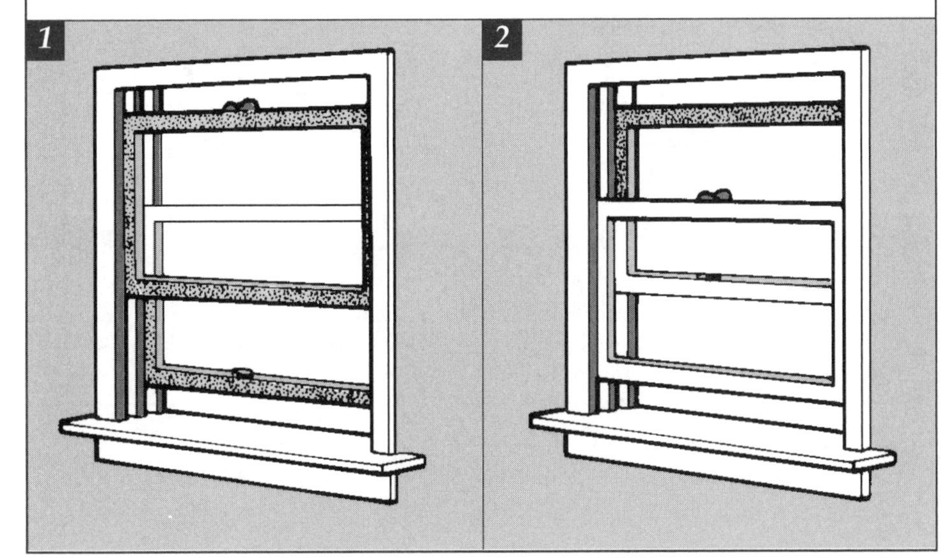

Painting Trim

A paint trimmer or edger is useful for painting clean edges along trim. This is especially helpful if you do not have a steady hand.

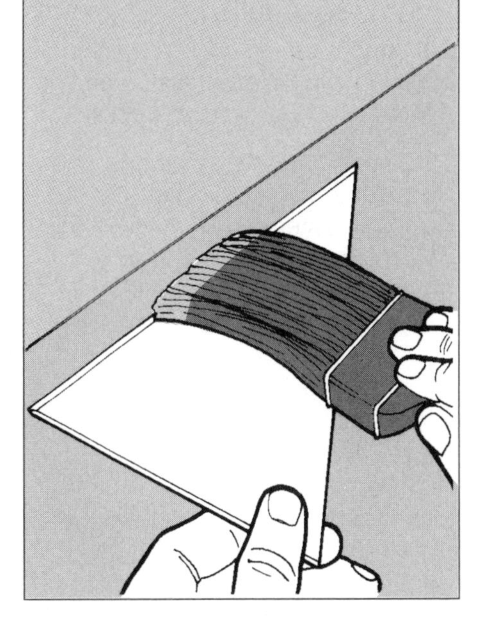

Wallcovering

Wallcovering is a somewhat new term that actually covers a host of different products, from traditional wallpaper to fabric-backed vinyl, paper-backed grass cloths, and even more exotic variations.

Several factors go into the choice of a wallcovering. Does the area to be covered get a lot of traffic? If it does, you should look for a covering that can withstand scuffs and can be cleaned easily—a solid vinyl covering for the highest hazard areas like kitchens and bathrooms, vinyl-coated papers for bedrooms and halls. A low traffic area can take a more delicate wallpaper. For added elegance, you can choose flocked wallpaper with its surface slightly raised in patterns of velvet-like synthetic fibers; paper-backed foils with a metallic surface creating a dramatic effect; or grass papers that provide a soft, neutral surface.

Some wallcovering is available pre-pasted; most must be pasted at home, sheet by sheet as it goes up. However it is pasted, wallcovering must be applied to a clean, smooth surface. No matter how attractive the covering, any blemish in the wall beneath will show through clearly and spoil the effect. It is best to strip walls of old wallcovering before applying new. In cases where the wall is not sufficiently smooth, wallpaper liner—a thicker, blank wall-covering —should be pasted on the wall as a foundation for a smooth surface.

To cover a wall you will need a bucket of water, a bucket for mixing paste, a wallcover smoothing brush—long and narrow with moderately stiff but pliable bristles—a long table, a long straightedge, scissors, a seam roller, a mat knife, a level and measuring tools.

Planning the Job

Wallcovering is sold in rolls of various widths. Because patterned wallcoverings must be matched side to side along the edge of the strips, there is a fair amount of waste in trimming to keep the pattern repeating properly. To estimate material needs, as a general rule, determine the number of square feet in the area to be covered, less openings like windows, doors and fireplaces. Then divide this by 30, a number derived by subtracting the likely wastage from the standard 36 square feet in a roll. Round up to the nearest whole number for ordering standard rolls. If you are buying other than standard 36-square-foot rolls, consult your dealer about how many you need.

The repeating pattern in wall-covering also requires careful planning of where the job should start and end.

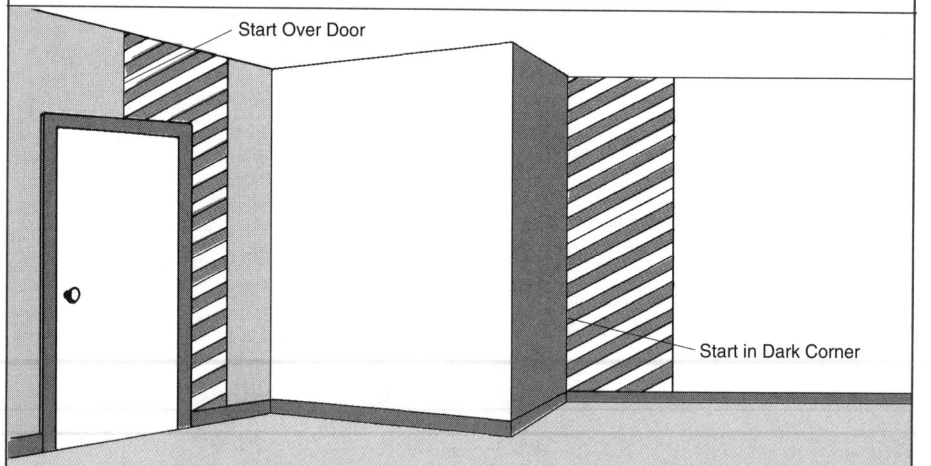

If you are covering an entire room, it is unlikely that the pattern of the covering will line up perfectly. Plan the job so separate pieces meet in the least conspicuous part of the room, such as over a door or in a dark corner.

Use a roll of wallcovering as a measuring stick to divide the wall into increments as wide as the paper. If the strips at either corner will be less than half a roll wide, begin in the middle of the room, shown left and right.

Cutting to Fit

1. Because wallcovering comes in rolls, it always must be cut to fit the height of the wall, and because full widths do not always fit exactly across a wall, it often has to be cut narrower along its length. To cut a piece to length, allow about 2 inches of overlap at the top and bottom to be trimmed off after the covering is on the wall. This lets you adjust a sheet up and down a little to meet the pattern properly.

2. Long cuts on wallcovering should be marked at both ends, measuring in from the edge that will meet the piece already on the wall. Long cuts are usually made to fit the wallcovering into corners and should be measured from the top and bottom of the wall because corners are rarely plumb.

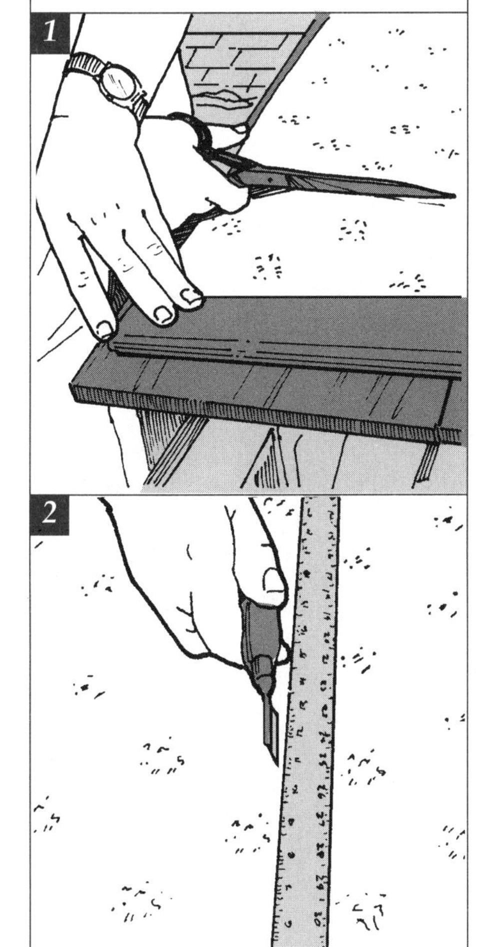

Pasting Wallcoverings

1 Mixing Paste. Wallcovering paste is available both premixed in liquid form and dry for mixing with water at home. If you are mixing your own, make it up about 30 minutes before you start.

2 Applying the Paste. Lay a piece of the covering that has been cut to length on your pasting table with one edge flush with a long edge. Paste the covering with a paste brush from the table edge to the middle and about half its length. Shift the covering across the table so that the other edge lines up along the other edge of the table and paste the rest of that side. Lining the covering up with the table edges prevents paste from getting on the table top.

3 Completing the Pasting. Fold the covering over on itself and pull the remainder up on the table to paste.

4 Folding the Strip. When the entire sheet is pasted, fold it into a manageable package that will be easy to carry to the wall. These packets can be set aside a few minutes to allow the paste to soften the covering.

1 If you mix your own paste, work the powder into the water until it has a smooth, somewhat viscous consistency.

2 Spread paste evenly with a wallpaper brush. Align the strip with the edge of the table to keep paste off.

3 Fold the pasted section of a strip over on itself (paste to paste) and paste the remaining section.

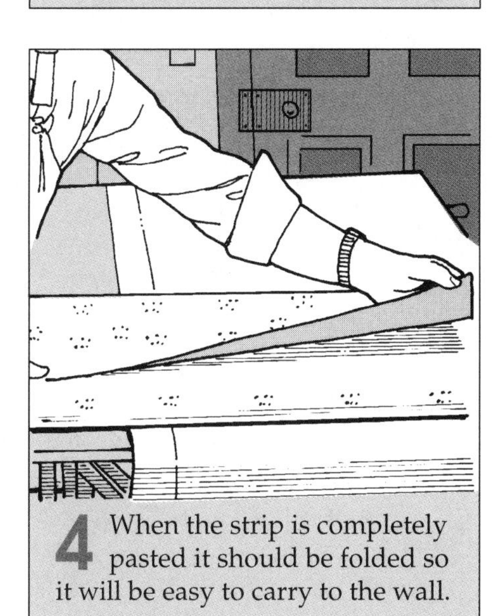

4 When the strip is completely pasted it should be folded so it will be easy to carry to the wall.

Putting Wallcovering on the Wall

1 Finding the Position. Decide at what point in the room you will start hanging the wallcovering and mark a vertical line at that point on the wall. This guideline will establish the position of subsequent sheets.

2 Carrying the Sheet to the Wall. Carry a pasted and folded piece of wallcovering to the wall, holding it with the top corners between thumb and forefinger, the rest of your hand supporting the rest of the sheet. This enables you to position the paper at the top, allowing the rest of the sheet to fall into place.

3 Positioning the Covering. Put the top of the paper against the ceiling, leaving a few inches of overlap, and shift it into position along the vertical guideline (in the drawing, the edge of a window that is plumb).

4 Brushing the Covering. Once the sheet is in position and laid reasonably flat by hand, use a wallcovering brush to smooth out wrinkles. Brushing action should be from the mid-line of the sheet toward the edges and toward the corners. Use the brush to tuck the paper into corners and along the ceiling line and at the bottom. Finish the surface by wiping it with a damp sponge.

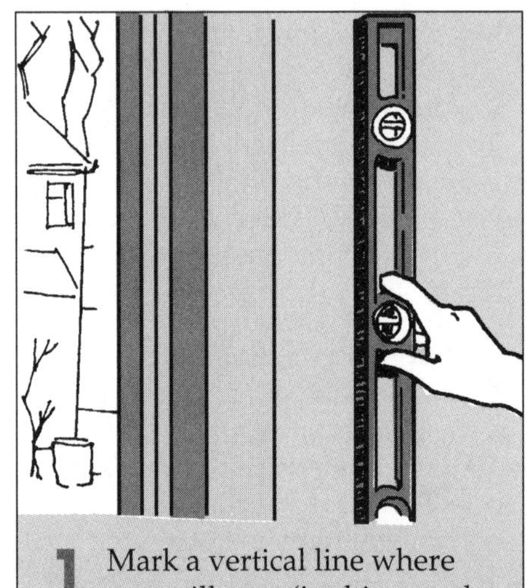

1 Mark a vertical line where you will start (in this case the starting point is along a window that is plumb).

2 Hold the covering at the top between thumb and forefinger so you can let it fall from the top.

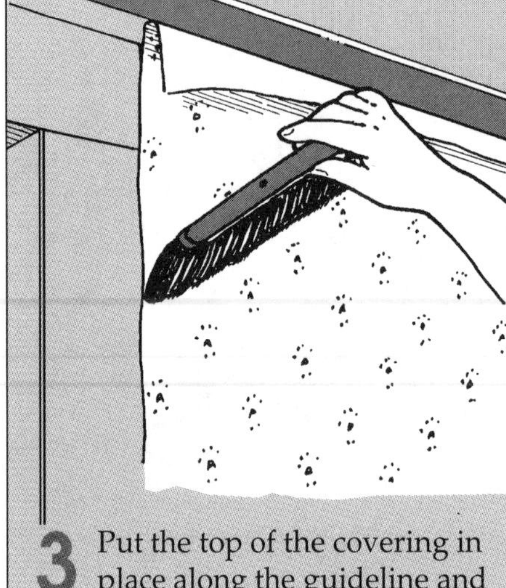

3 Put the top of the covering in place along the guideline and let the rest of the strip drop down the wall.

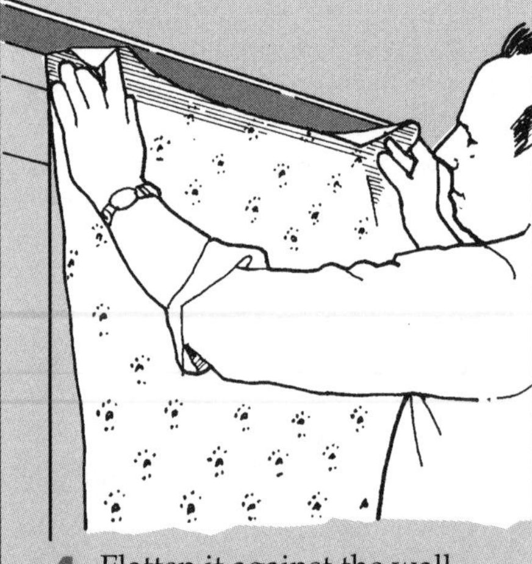

4 Flatten it against the wall with a wallcovering brush, stroking from the center out.

Trimming for an Exact Fit

1. When the covering is hung and brushed out, mark the line where it meets the ceiling, floor, or corner by dragging the back of a scissors blade along the joinder. Peel the covering down enough to permit cutting, cut along the line, and stick the covering back against the wall, brushing it down.

2. At windows, hang a sheet over the window that must be cut around. Notch the corners back to the edge of the window, then score and cut.

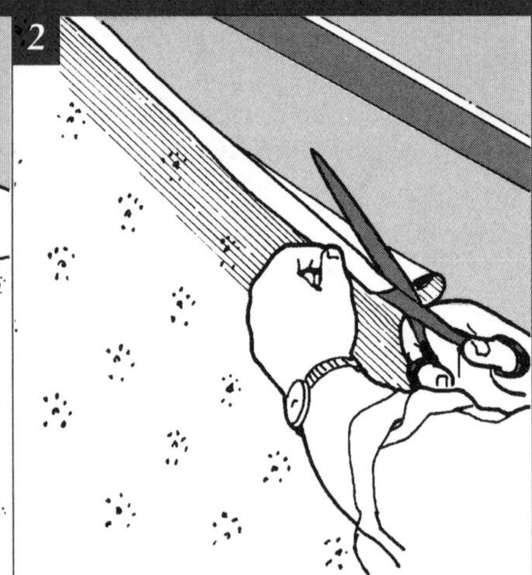

Hanging Tight Seams

1. Position a second strip along the edge of the first so that the pattern lines up and the edges of the sheets are butted together tightly—not overlapped or pulled apart at all—with a very slight ridge at the junction. This ridge will subside.

2. After the paste has started to dry and the edges have sunk back to the wall, use a seam roller to flatten the seam and press the edges of the sheets firmly into the paste. Roll once up and down. Do not roll over the seam again and again for you may create a shiny track on the covering.

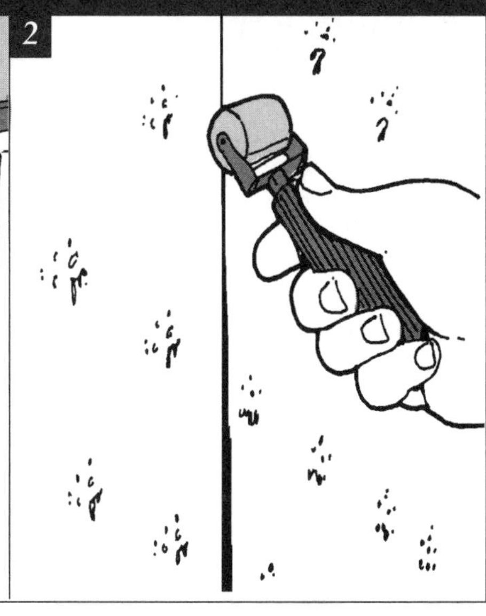

Turning Inside Corners

1 Hanging the First Strip. Corners are rarely straight in any but a brand new house, so wallpaper must be cut and fitted to carry the pattern. Measure from the edge of the sheet before a corner, to the corner at the top of the wall and the bottom of the wall. Add 1/2 inch to these measurements and transfer the larger of the two to a sheet and cut it lengthwise. Hang the sheet against the edge of the previous sheet, letting the other edge turn the corner. Brush the sheet out and tuck it into the corner.

2 Marking for the Second Strip. Measure the width of the remaining section of covering and subtract 1/2 inch. Transfer this measurement to the corner of the uncovered wall and use a plumb bob to find the vertical line that touches this measurement. Mark this vertical as a guide to hanging the next piece.

3 Hanging the Second Strip. Hang the sheet, positioning it against the line and brushing out as usual. Use the brush to tuck in the edge that meets the corner.

4 Fitting the Second Strip. If the second sheet overlaps the corner, score it with scissors, peel it back slightly, cut along scored line.

1 Measure from the last sheet to the corner at top and bottom, add 1/2 in. and cut a strip to that width.

2 Mark a vertical line a distance from the corner equal to the width of the remainder of the strip.

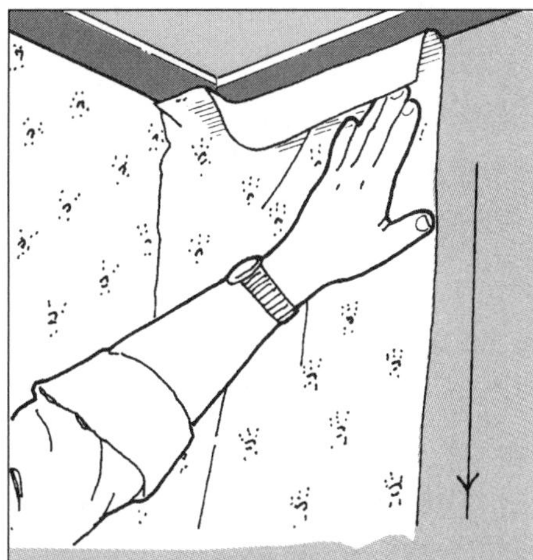

3 Use the vertical guideline to position the second strip. Brush the overlap into the corner.

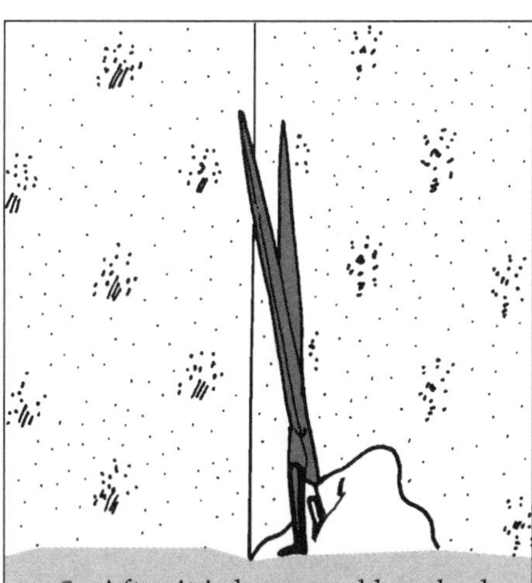

4 After it is hung and brushed out, score the second strip in the corner and cut it to fit.

Hanging Objects Permanently

The right way to hang anything on a wall depends on two things: the construction of the wall and the weight and size of the object.

For general purposes, walls can be divided into two categories. They are either hollow (wallboard, plaster, or other surfaces over a stud frame) or solid (concrete block, brick, or plaster over masonry). Attachments to hollow walls are usually made with (a) nails into the wall surface or through it into a stud behind, (b) an expansion anchor or toggle bolt, both of which are mounted through holes in the surface and grip the wall surface from behind, or (c) a plastic anchor that is a sleeve fitted snugly into a drilled hole to hold the wall with friction when it is expanded by a screw driven into it.

Attachments to solid walls are made either by driving a masonry nail into the wall or by inserting a plug or wall anchor of some kind into a drilled hole. The plug or anchor in turn accepts a screw. The drawings below and right show the most common wall attachments. There are a great variety of different fasteners suitable for different jobs—from hanging a small picture to installing heavy kitchen cabinets—and different wall conditions from the basement to the attic.

Very Light Weight. Adhesive picture hooks have adhesive on the back. Before applying, be sure the wall is completely clean so adhesive will stick.

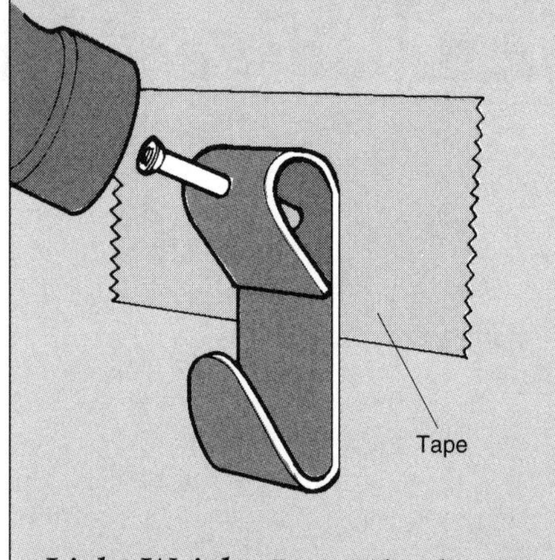

Light Weight. Picture hooks can be nailed to a hollow wall between studs to hold lightweight items, up to about 20 pounds. Tape protects surface from cracking.

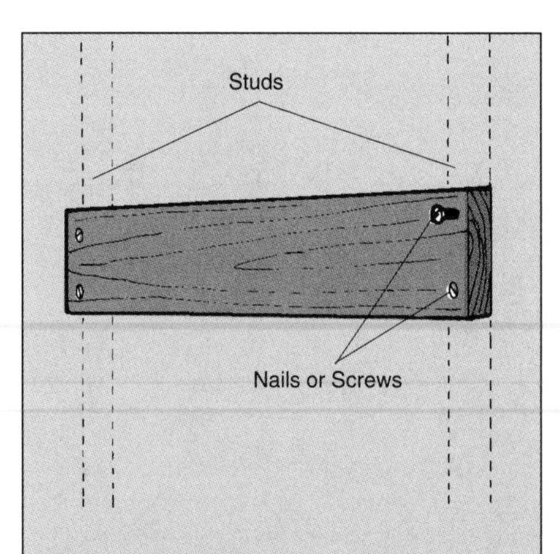

Heavy Weight. Such heavy objects as cabinets should be attached to a brace that has been securely screwed or nailed between two studs.

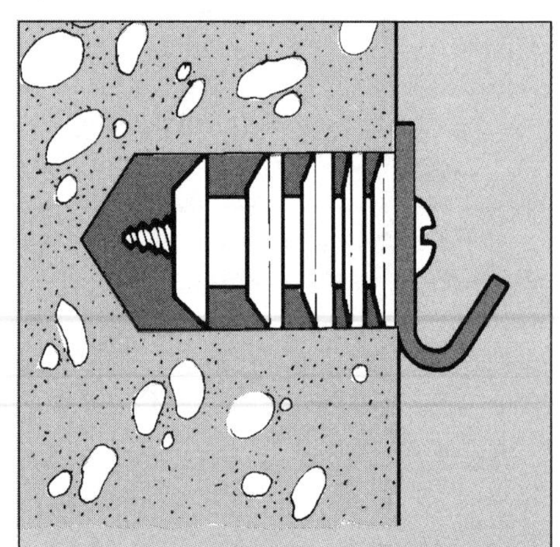

Solid Wall, Heavy Weight. Anchors and plugs should be seated in holes drilled to their diameter so the fit is snug. A screw then jams the plug into the wall.

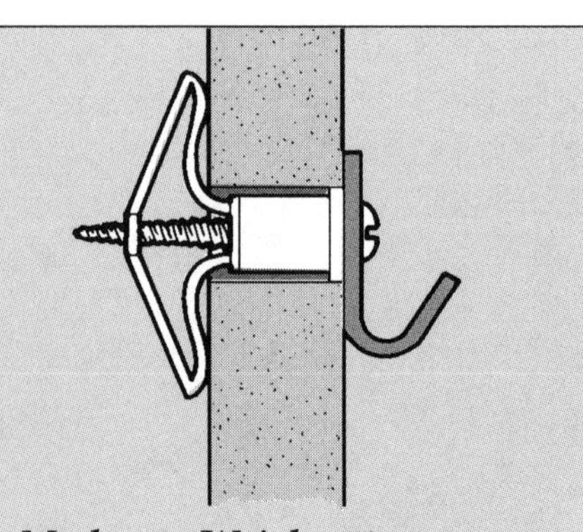

Moderate Weight. Plastic wing anchors work like expansion anchors. The wings are held closed to insert; they open as the screw is driven.

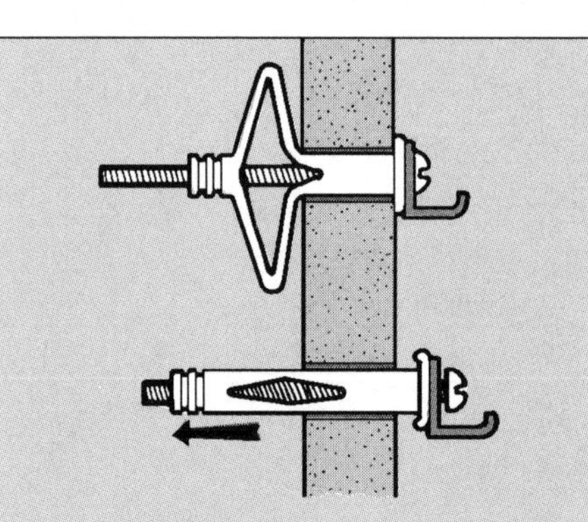

Moderate Weight. A hollow wall expansion anchor opens when the bolt is fully tight. It must be seated in a hole the diameter of the anchor.

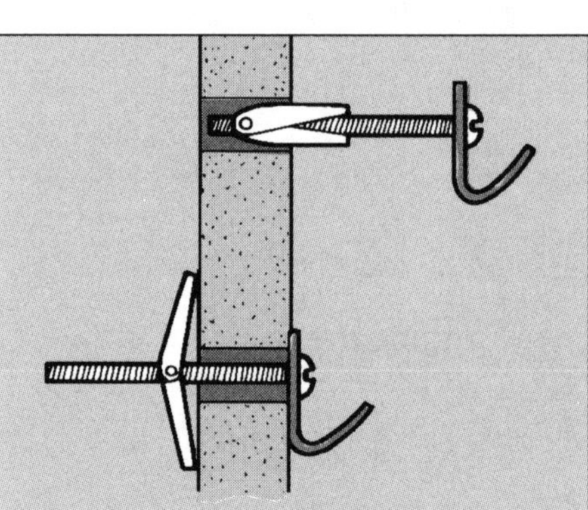

Moderate Weight. Toggle bolts have wings under spring tension that are held closed as the bolt is inserted in the hole in the wall, then pop open.

CEILINGS

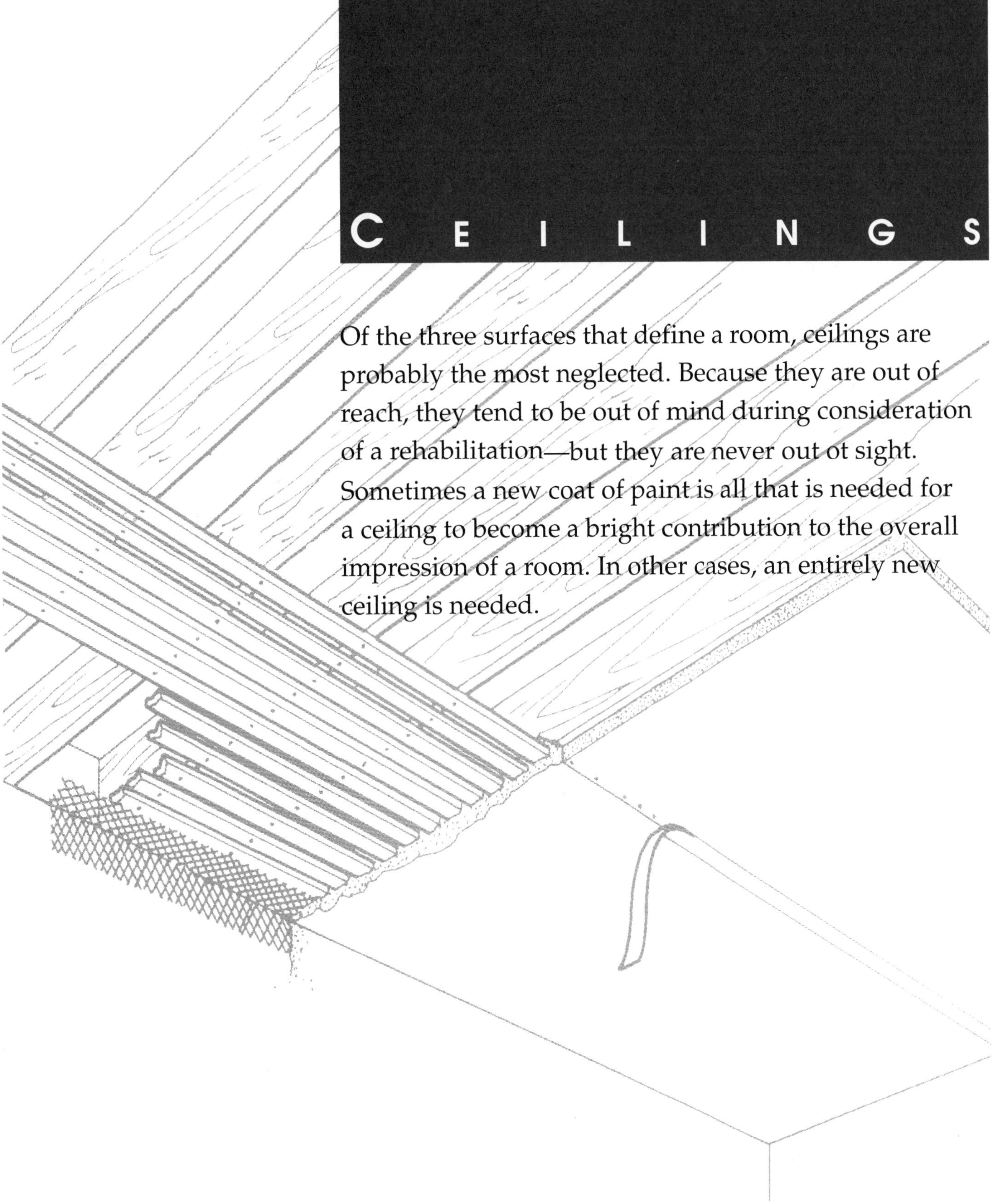

Of the three surfaces that define a room, ceilings are probably the most neglected. Because they are out of reach, they tend to be out of mind during consideration of a rehabilitation—but they are never out ot sight. Sometimes a new coat of paint is all that is needed for a ceiling to become a bright contribution to the overall impression of a room. In other cases, an entirely new ceiling is needed.

Anatomy of Ceilings

Structure

In a sense, an ordinary plaster or wallboard ceiling has no structure—it is merely a surface attached to the bottom of the floor above. Another way to consider the structure of a ceiling is that it is the same as the structure of a floor. Either lath and plaster or wallboard is attached to the undersides of the joists that tie the walls of the house together. As a house settles, these joists may slip slightly out of line causing an uneven ceiling. This problem is most easily corrected by installing a new surface onto furring strips that have been shimmed level, or by suspending a ceiling on a grid system hanging from the original ceiling. Both procedures are discussed below.

The inside of a ceiling is most important in any lighting installation because joist location often determines where you can install a fixture; bringing power to the fixture usually requires running wiring between joists. The job is easiest when you can work in an attic or crawl space that has no flooring at all or only rough flooring that can be removed and put back without harm.

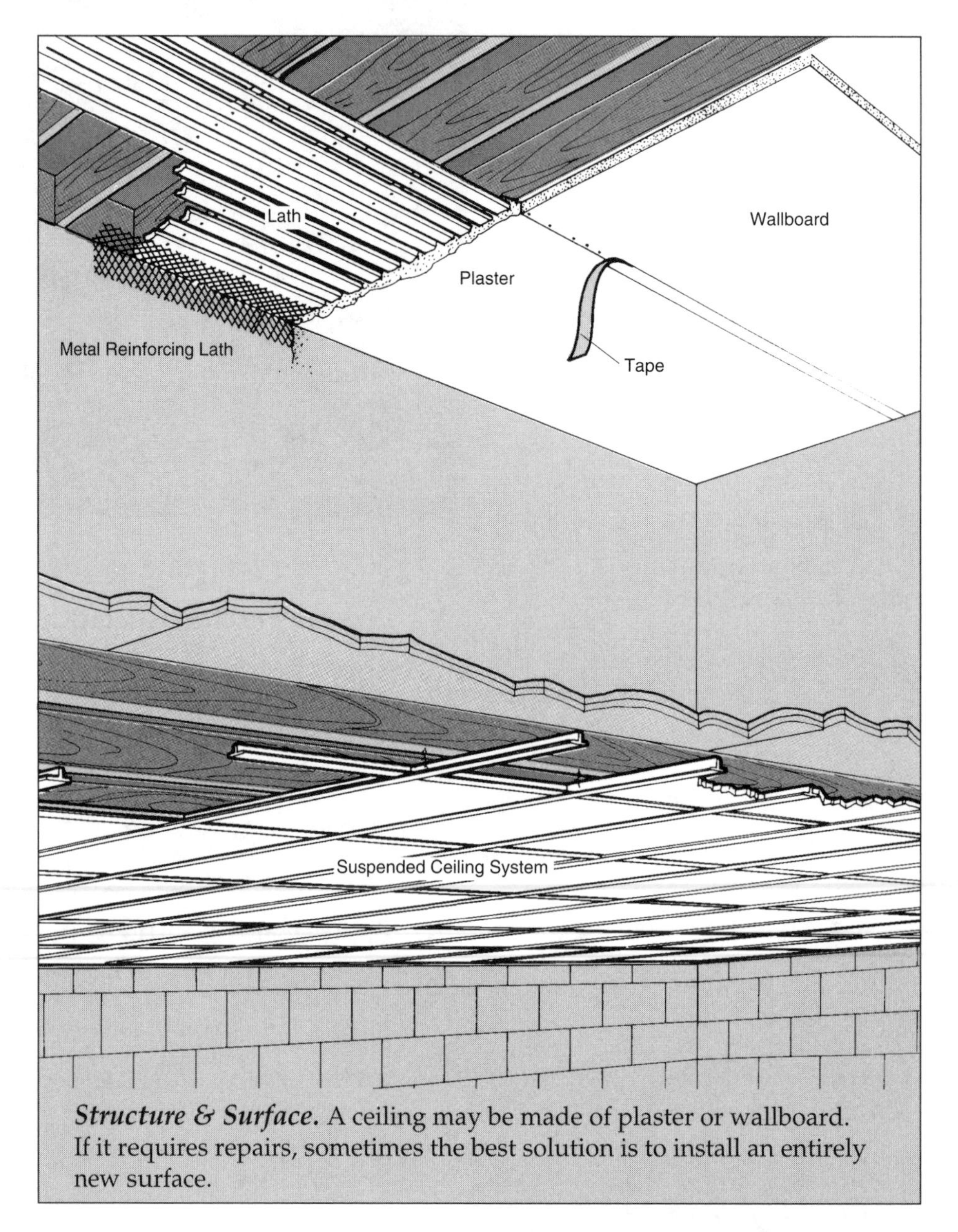

Structure & Surface. A ceiling may be made of plaster or wallboard. If it requires repairs, sometimes the best solution is to install an entirely new surface.

Surface

You will be concerned with ceiling surfaces in new construction, when a ceiling is in bad shape, or when the ceiling has never been finished, as in an unfinished basement or attic.

On an unfinished ceiling, wallboard is simply attached to the underside of the joists with adhesive and nails, then the seams are taped and feathered like seams on a wall. The only difference between covering a ceiling and a wall with wallboard is that working above your head is much more difficult. If you are covering a finished ceiling with wallboard to conceal defects, you may need to level the existing ceiling with shimmed furring strips first. Again this work is essentially the same as preparing walls for new surfaces.

Ceiling tile, often referred to as acoustic tile, now comes in a wide variety of colors and finishes. The tiles are far easier to manage than sheets of wallboard.

Ceiling-mounted light fixtures often require opening the ceiling, working inside, then closing it up. The most extensive intrusion into the ceiling you can make is to open it right through to daylight. Skylight installations involve a lot of work, but the effect of natural light overhead in a room makes the effort worthwhile.

Caution

Always have a plaster sample tested for asbestos content before removing old plaster walls.
If asbestos is found, contact a licensed abatement contractor regarding the removal and disposal of hazardous building materials.

Typically the removal of asbestos requires not only a special vacuum with filters that remove dust from surfaces, but special "air scrubbers" that remove dust from the air as well. The person doing the work must wear a respirator which is connected to an outside air supply.

Repairing Ceilings

Ceiling repairs are similar to wall repairs, but they are easier to accomplish if you have access to the ceiling through a crawl space or an unfinished (or roughly finished) attic. If a ceiling is stained by leaks but is otherwise undamaged, let the ceiling dry. Then prime it to hide the stain before repainting.

Small Holes in Wallboard

From above, set a piece of wire mesh or lath overlapping the hole and fix the corners to the wallboard with joint compound. With the mesh secured, fill the hole from below with plaster, building up layers, and finish with joint compound smoothed with a taping knife (see top left).

Large Holes in Wallboard

If you have access from above, cut the damaged wallboard back to the edge of the joists on either side of the damage and frame the opening with 1x2 (see top right). Secure the wallboard patch, then tape and feather the seams.

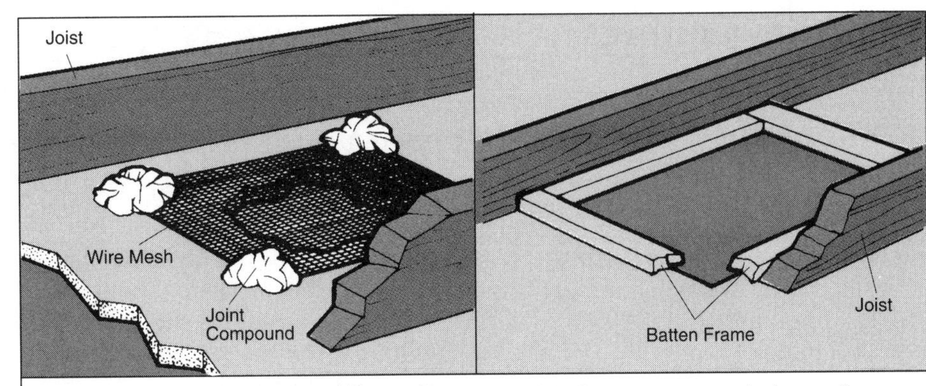

Repairing Holes in Wallboard. For small holes, secure mesh from above with joint compound, then fill with plaster and finish with compound (left). If you can't work from above, use method shown on page 8. For large holes, nail a batten frame to joists so the bottom is flush with bottom of joists (right). Then secure wallboard patch to frame.

Fixing Bulging Plaster

If you have access from above, prop up the bulge with a flat surface larger than the bulge, carefully pull and scrape out plaster protruding between the laths, wet the old plaster, and fill between laths—forming new ridges over them—with new quick-setting plaster from above.

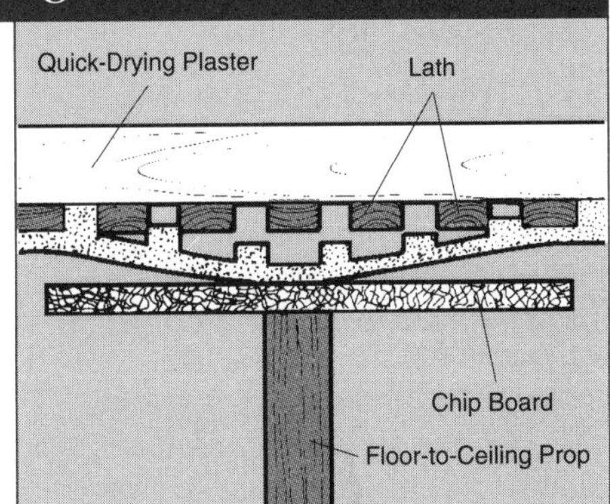

Small Holes in Plaster

1. From below, either tack metal lath to the existing lath or stuff the hole with newspaper soaked in plaster.

2. Fill hole with plaster to within 1/4 inch of the surface and let it set and recede before finishing.

3. Wet the patch and apply a last coat of finish plaster, smoothing the surface with a straightedge.

1

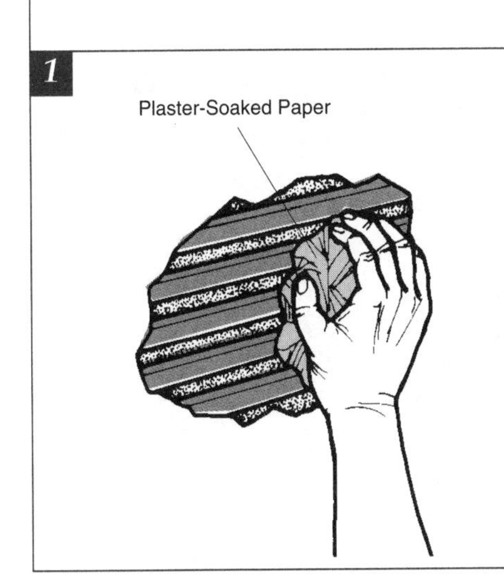

2

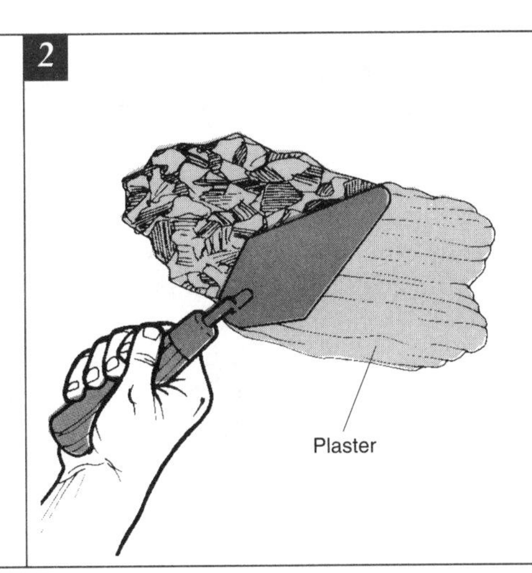

3

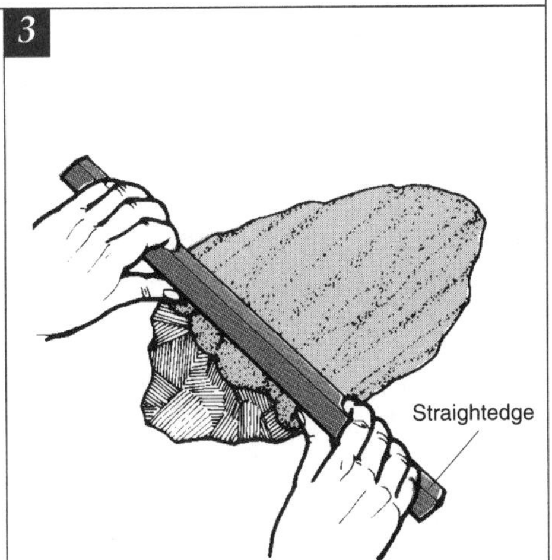

Finishing Ceilings with Wallboard

There are a variety of materials for finishing unfinished ceilings or covering old ceilings with new surfaces. Directions for installing the three most common—wallboard, acoustic tile and a suspended panel ceiling—are presented in the following pages.

Each way of finishing a ceiling has its own advantages and you should choose your approach after considering all three. Wallboard offers a ceiling surface that is smooth and can be painted, wallpapered or paneled to harmonize with the rest of the room. It also is the best choice if soundproofing the ceiling is a concern (see page 64).

Putting wallboard on a ceiling is a demanding job because it requires raising and holding the sheets, nailing, and finally taping them all overhead at inconvenient angles. Acoustic tile (see page 58) which is stapled to furring strips on the ceiling, or is applied to a finished ceiling with adhesive, is available in various textures and hues. However, it offers limited decorating possibilities and is usually found in basement recreation rooms and shops, home offices and attic conversions. Tile is the easiest of the three to install. Suspended ceilings (see pages 60-62) have the same properties and about the same appearance as acoustic tile but are hung on a grid of metal suspended from the ceiling. This system allows you to hide pipes, ducts and electrical fixtures that run below the bottom of ceiling joists and also permits the installation of lighting to shine through translucent panels in the ceiling.

On an unfinished ceiling, wallboard is nailed to the exposed joists. To refinish a ceiling with wallboard, you can nail the wallboard directly to the ceiling (if it is level and you can locate the joists to nail into). If a finished ceiling is not level, attach shimmed furring (see page 58) before installing wallboard. Finish as you would a wall.

Planning a Wallboard Ceiling

Put whole sheets along walls and trim pieces in the middle. The seams between the ends of sheets should be staggered to facilitate taping. Where whole sheets lie side by side, start the second row with the piece that was trimmed off the end of the first, if this remainder is longer than 18 in. If the sheets fit along a ceiling without enough overlap to start the next row, start that row with a 6-foot piece and trim with the remainder.

Hammering Over Head

Grip the hammer with your thumb along the handle and swing it from the elbow.

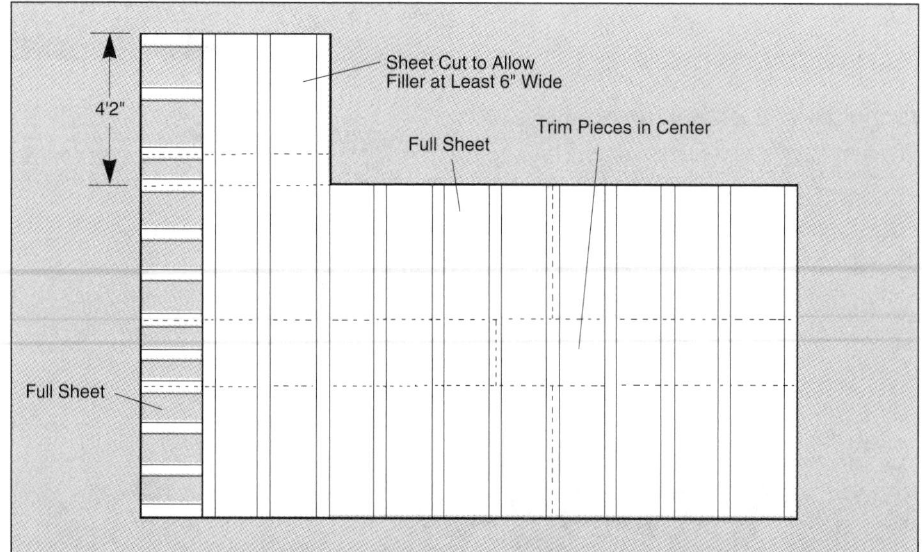

Planning a Wallboard Ceiling. Sketch a layout for wallboard sheets before making any cuts to be sure you don't encounter awkward piecing. If you must cover an area only a few inches wider than a sheet, cut down the sheet (as in the alcove shown) to allow a trim piece at least 1 ft. wide.

Providing Nailers

Check along the walls that run parallel to the joists to see whether you have a nailing surface to which you can fix the ends of the sheets. If there is none, as is often the case on end walls in basements where the last joist may sit on the middle of the top plate, attach nailers.

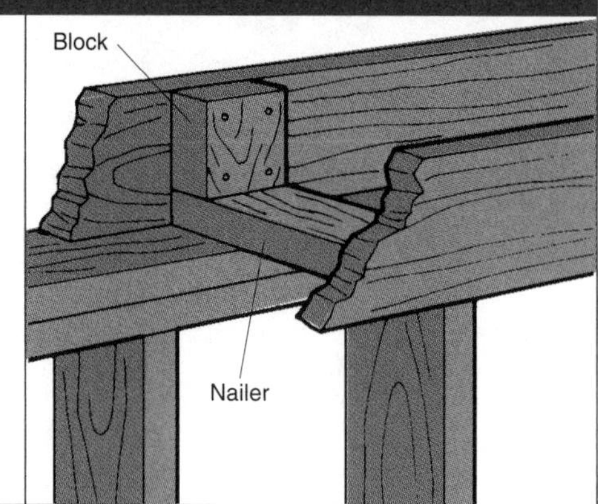

Installing Wallboard Ceilings

1 Marking the Joists. Make a layout plan as described on the previous page. Mark the locations of the joists below the ceiling level to guide you when nailing wallboard.

2 Applying Adhesive. To minimize the number of nails you have to drive and plaster over, use adhesive along the ends of the joists to help hold the wallboard. Run beads of adhesive along joists, squiggling the bead where sheets will butt so both ends will press into adhesive.

3 Nailing Wallboard. Build a T brace from a piece of 2x3 with a piece of 1x4 nailed across the top. The 1x4 should be nailed straight across with its top edge slightly raised above the end of the 2x3 to prevent it from gouging the wall-board. The entire brace should be 3/4 inch longer than the height of the new ceiling so it can be wedged against the floor for a tight fit. Lift the wallboard into place and have a helper prop it up with the brace. Nail around the edges and across the joists at 16-inch intervals with $1\frac{3}{8}$-inch ringed wallboard nails.

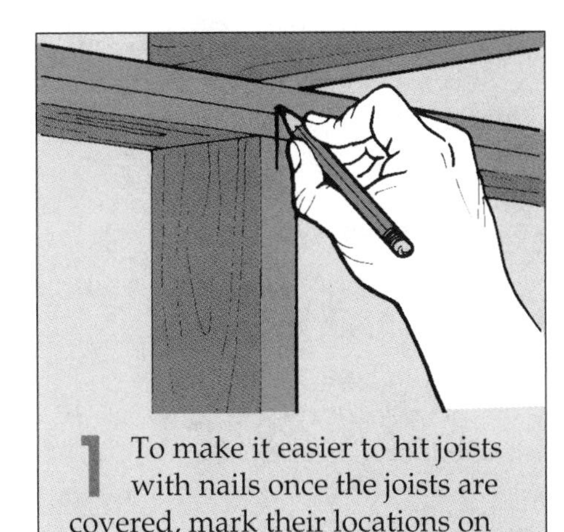

1 To make it easier to hit joists with nails once the joists are covered, mark their locations on the studs or wall.

2 Run straight beads of adhesive along joists to hold sheets; wavy beads along joists where sheets will butt.

3 Raise the sheet to the ceiling and have a helper brace it in position with a homemade T brace. Nailing will be easier if you use a platform of planks between the lower steps of two ladders or a ladder and a sawhorse.

Single-Handed Installation

1. Attach a 1x4 longer than 4 feet to the wall, 1/2 inch below the ceiling where the end of the first sheet will go—or 3/8 inch below depending on the thickness of the wallboard. Place a step-ladder about 6 feet away in the direction the sheet will sit against the ceiling. Rest one end of the sheet against the wall and put the other end on the top of the ladder. Raise the end at the wall into position on top of the 1x4 strip. Then, holding the T brace with one hand, raise the other end of the sheet to the ceiling with your free hand and slide the brace into position to support it. Check the position of the sheet before nailing it.

2. After nailing the first sheet, attach a 1x4 brace at the open end. Move the ladder and repeat the procedure, seating the second sheet between the 1x4 brace and the joists, and supporting the other end with the T brace.

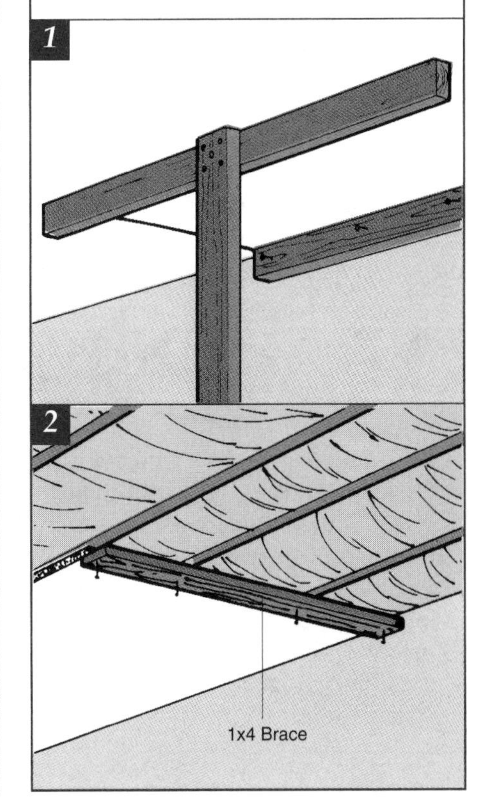

Installing an Acoustic Tile Ceiling

Acoustic ceiling tile or panels in a suspended ceiling (see pages 60-62) tend to give a room either a business-like or informal character and is appropriate in a basement or recreation room. A property that makes ceiling tile desirable wherever there is much activity is its ability to absorb sound within a room. Wallboard, provides greater soundproofing between rooms.

Not all acoustic tile looks like the familiar kind used in schools. It is available finished with decorative surfaces and in a variety of light colors. Standard sizes are 12-inch squares and 12x24-inch rectangles, and larger, tongued and grooved to fit together. Not all tile is sound-absorbing so read labels carefully—tile with acoustic properties is always marked as such.

Ceiling tile can be installed with adhesive directly to clean, smooth ceilings. If a ceiling is rough, not uniformly level, or unfinished, you must put up furring strips before installing tile. Furring a ceiling is much like furring a wall, although the placement of strips is different. Tile is stapled to the furring.

To fur a ceiling, you will need a long, straight piece of lumber for leveling across the strips, enough 1x3 to cover the ceiling according to your plan, and measuring and light carpentry tools. When installing the tile, you will need a utility knife, a metal straightedge, and a heavy-duty staple gun to drive 1/2-inch staples.

Furring a Ceiling

1 Adding Nailers. Inspect the edges of the ceiling along the walls that run parallel to the joists—it is often necessary to add a nailing surface there. This surface must be continuous along the length of the wall and can most easily be added by running a 2x8 along the top of the wall, with its bottom edge flush with the bottom of the joists (or other width depending on how far the joist above the wall is offset from the edge of the wall). Fix it in place with blocking.

2 Shimming the Furring. Make a sketch of the ceiling and position the furring according to the kind of surface you are applying: The strips should be 12 inches for ceiling tile and 24 inches for wallboard installed on center. For tiling, measure the distance from the midpoint of the walls to the corners in 1-foot intervals. If the distance from the last whole foot to the wall is greater than 3 inches, center a furring strip at the midpoint; if less than 3 inches, put furring strips 6 inches on center on either side of the midpoint. Snap a chalk line across the ceiling or the bottoms of the joists, perpendicular to the joists, as guidelines for attaching the furring. If the ceiling is not level, as is usually the case, find the lowest point and attach a furring strip to the nearest intersection of a joist and a guideline. Use a level to shim this strip along the ceiling to the walls on either side.

3 Checking for Level. Use a long level or a level taped to a straight board to check that subsequent strips are level.

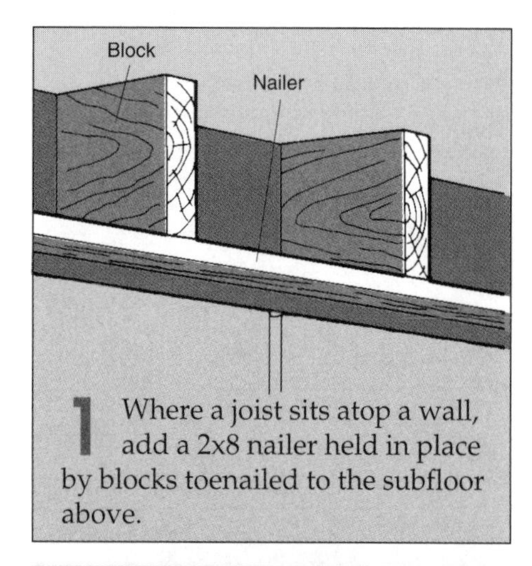

1 Where a joist sits atop a wall, add a 2x8 nailer held in place by blocks toenailed to the subfloor above.

2 Attach the first piece of furring to the lowest point on the ceiling and shim the others to that level.

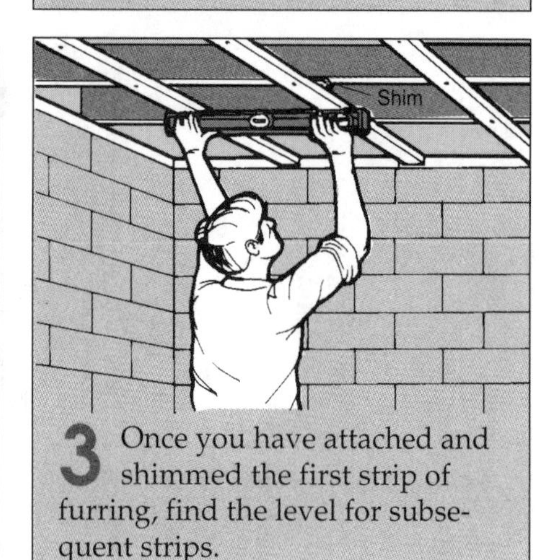

3 Once you have attached and shimmed the first strip of furring, find the level for subsequent strips.

Double Furring

Where pipes, wiring or other obstructions hang 1/2 inch below the joists, furring will cover them. If they hang up to $1^{1}/_{2}$ inch below the joists, double the furring by installing a first layer across the joists at 24-inch intervals, then install a second layer perpendicular to the first with the spacing required by the tiles.

Installing Tile

1 Marking Guidelines. Mark the distances from adjacent walls that equal the trim sizes of the tiles that will go around the margin of the ceiling. The line that runs parallel with the furring will fall on the center of the furring strip nearest the wall.

2 Installing the First Tile. Position the first tile inside the intersection of the guidelines with the grooved edges facing the adjacent walls. Staple the tile in place along the tongues with a staple gun and 1/2-inch staples. Continue, fitting the tongues and grooves together, until the ceiling of full tiles is complete.

3 Trimming Tiles. Measure the distance between the edge of the full tiles and the wall and transfer this to a tile, measuring from the edge of the surface toward the grooved edge. Cut off the excess on the grooved edge and put the tile in place with the cut edge to the wall.

4 Measuring for a Cutout. To measure for cutouts in a tile, put the tile in place, with tongue in groove. At the center of the obstruction, mark the edge lightly (if you mark the face it will show). Reposition the tile and do the same on the adjacent face. Use a square to mark the intersection of these lines on the face of the tile. This is the center of the area to be cut out. Outline the shape with measurements taken from the obstruction and cut.

5 Finishing with Molding. Finish the ceiling with cove molding along the edge to conceal the cut edges of the tile where they butt against the walls.

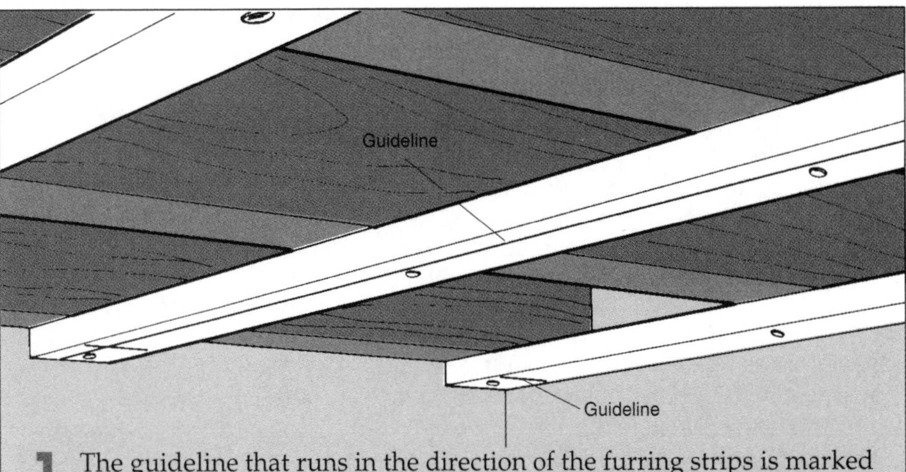

1 The guideline that runs in the direction of the furring strips is marked along the center of the first strip away from the wall. The guideline running perpendicular to the strips should be marked between the points indicating the last full tile width.

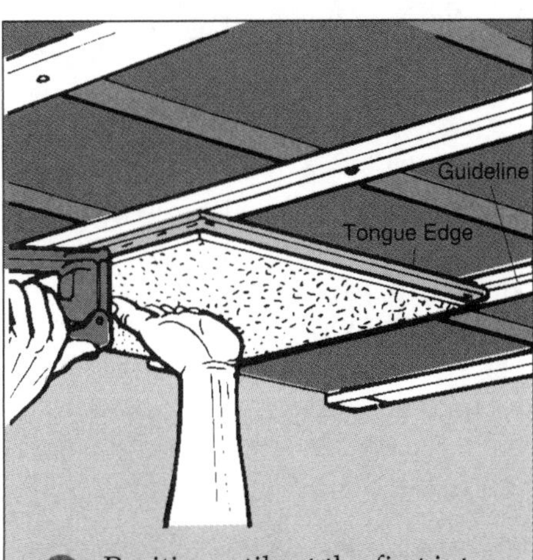

2 Position a tile at the first intersection of guidelines with its tongue edges facing away from the walls.

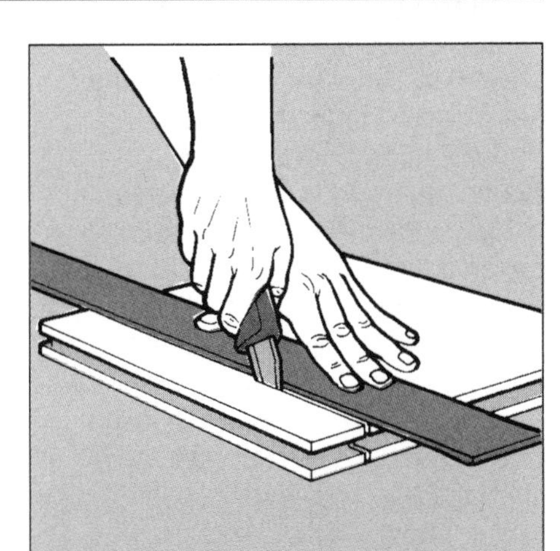

3 Tiles for the edges of the ceiling should be trimmed along the grooved edge (or edges for corner tiles).

4 Seat the tile and mark the edge at the center of the obstruction; repeat on the adjacent side.

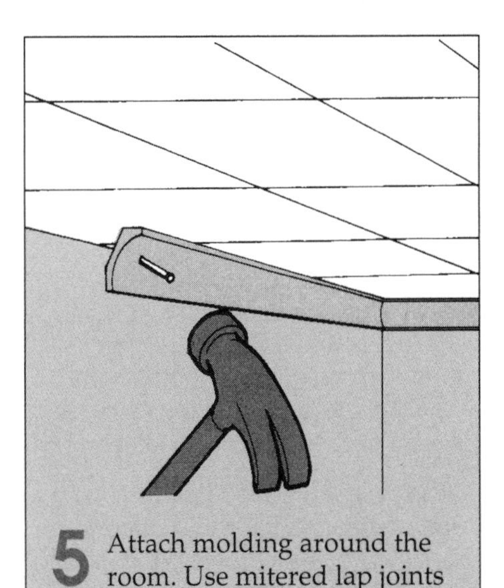

5 Attach molding around the room. Use mitered lap joints to make up required lengths.

Installing a Suspended Ceiling

An easy, relatively inexpensive way to finish an unfinished ceiling, or renew an old one, is to suspend a ceiling of lightweight panels on a grid of metal channels which hangs on wires from the original ceiling.

Such a system offers several practical advantages as well as a clean modern look. Because they hang below the level of the original ceiling the suspended panels hide pipes, duct work, and electrical connections, even when they project from the existing ceiling. Pipes that project lower than the height you choose for the ceiling can be boxed in with paneling (see page 63). The panels in the ceiling have good sound absorption properties; they swallow sound bouncing around inside a room, although they will not keep sound from passing through the ceiling to rooms above. A suspended ceiling is leveled as it is installed, without the need for any correction in the original ceiling above. The panels simply sit in the grid without attachment and can be lifted out at any time for access to the space above. Finally, suspended ceilings provide an easy way to illuminate a room from above. This can be done with luminous panels in which fluorescent tubes have been installed or with fluorescent light tubes attached to the original ceiling shining through translucent plastic panels fitted into the grid system instead of the usual fiberboard panels. Such lighting can be installed in a few places or the entire ceiling can be made a light source (see page 63) with fluorescent tubes.

The different parts of a suspended ceiling are: wall angle, cross Ts, runners, (shaped metal components that interlock to hold up the ceiling panels), and the panels themselves.

To put up a suspended ceiling, you need measuring tools, including a level and chalk line, a hammer, and a pair of heavy pliers to bend the suspension wires.

1 Planning the Job. Panels for a suspended ceiling are available in 2x2-foot and 2x4-foot sizes. The latter works best if you are placing fluorescent lighting above because it fits the standard fluorescent tube length. Smaller panels require twice as many cross Ts to hold them up and there are twice as many panels to install, so the job will take a little longer. Wall angle and runners are sold in different lengths up to 12 feet and can be overlapped to span greater distances. Cross Ts are 2 feet long.

2 Getting the Hardware. The easiest way to determine the amount of hardware and paneling needed is to make an installation plan before you buy the materials. This will let you count the number of feet of wall angles, runners, and cross Ts and the number of panels you will need. It will also give you the opportunity to plan for lighting and make decisions about how the panels will fit, before you get started. Draw the outline of the room to scale on graph paper and mark the position of the runners and cross Ts.

3 Fitting Cross Ts. To determine a symmetrical layout of panels that avoids the need for radically narrower panels to fill in along the edges, measure two adjacent walls and find their midpoints. Measure from the midpoint of the wall parallel to the joists; the runners will be attached perpendicular to this wall. If the distance from the last whole foot of the measurement is less than 6 inches, runners should be placed exactly 1 foot on either side of the midpoint; if it is more than 6 inches, a runner should be installed at the midpoint. Do the same calculation along the other wall, this time using 4-foot intervals. If the last full 4-foot interval ends $3^1/_2$ feet or closer to the end of the wall, put a line of cross Ts on the midpoint. If it is farther than $3^1/_2$ feet, put the first line of cross Ts exactly 2 feet on either side of the midpoint.

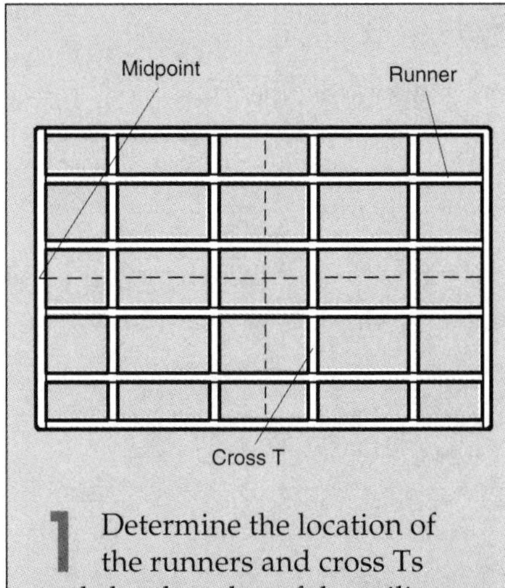

1 Determine the location of the runners and cross Ts and sketch a plan of the ceiling to determine material needs.

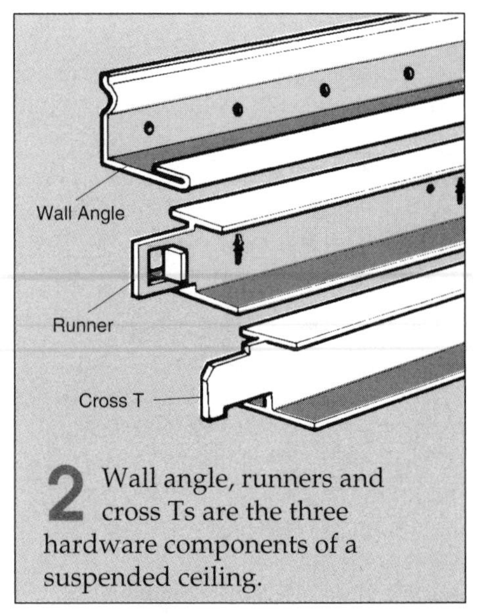

2 Wall angle, runners and cross Ts are the three hardware components of a suspended ceiling.

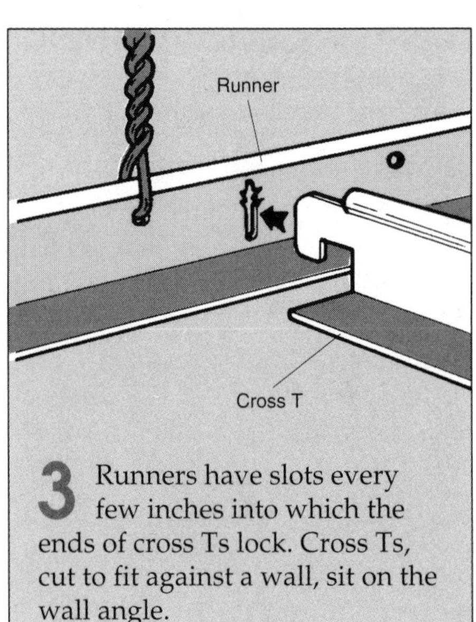

3 Runners have slots every few inches into which the ends of cross Ts lock. Cross Ts, cut to fit against a wall, sit on the wall angle.

4 Marking the Ceiling Height. After planning a layout, measure up from the floor to the height where you want the ceiling. The standard height is 7 feet, 6 inches, which is also considered a minimum height for lighting in a suspended ceiling. The ceiling should be no closer than 3 inches to any projection. You can hang it a few inches lower to avoid projections (but not too low for comfort) or box them in (see page 63). Mark the ceiling height on all four walls and stretch butcher's twine between nails tacked at that height on opposite walls, to check for horizontal with a string level. Make any adjustments necessary, then use a chalk line to mark the height around the room, using a level to keep it horizontal.

5 Attaching the Wall Angle. Nail or screw lengths of wall angle around the perimeter of the room at the ceiling height line. Butt pieces of wall angle to span long distances and overlap pieces at the corners. Cut pieces to fit with tin shears and remember to account for the fraction of an inch thickness of the wall angle when cutting to butt against a piece on an adjacent wall.

6 Marking for the Runners. Find the midpoints of the wall and snap chalk lines across the joists (or in either direction across a finished ceiling) at 2-foot intervals, starting at the midpoint of the wall or at a point 1 foot to either side, depending on the layout plan.

7 Stretching Strings to Mark Cross Ts. Mark the locations of the cross Ts on the walls parallel to the runners at 2-foot or 4-foot intervals, depending on the size of your panels. These lines should start at the midpoint or 2 feet off center, according to your layout. Attach strings across the room from nails tacked at the bottom of the wall angle at its intersections with the cross Ts. These mark the height at which you will hang the runners.

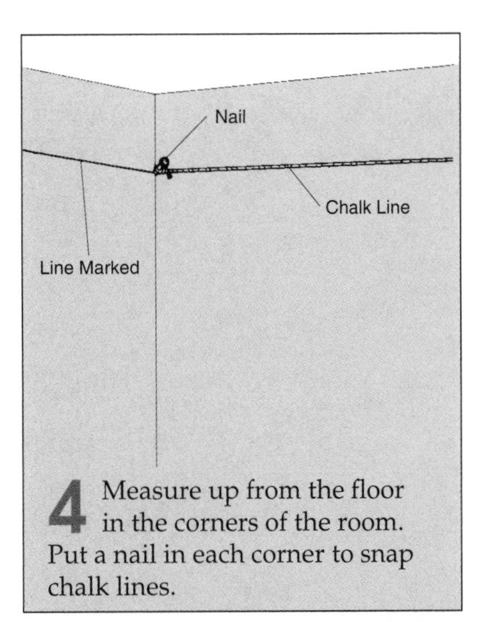

4 Measure up from the floor in the corners of the room. Put a nail in each corner to snap chalk lines.

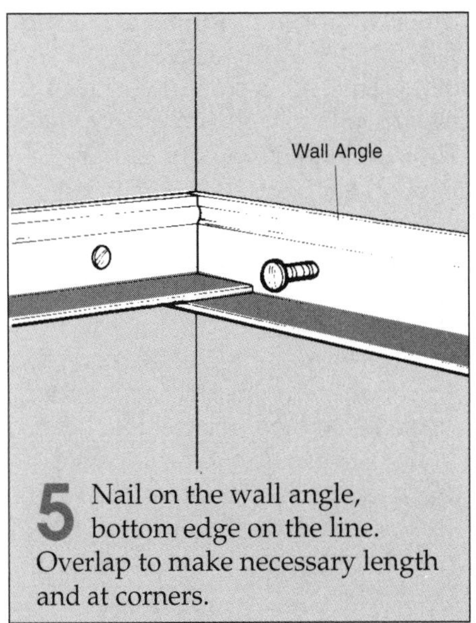

5 Nail on the wall angle, bottom edge on the line. Overlap to make necessary length and at corners.

6 Stretch a chalk line between nails at the midpoint (or 1 ft. to either side, depending on your layout) perpendicular to the joists and snap line(s). Repeat at 2 ft. intervals across the ceiling. These mark the location of the runners.

7 At the points on the adjacent walls where the cross Ts are to be located, stretch strings from nails driven into the wall exactly at the bottom of the wall angle. These mark locations of cross Ts as well as the level of suspended runners.

8 Hanging the Runners. Starting with the joists at either end of the ceiling, put a screw eye into every fourth joist at each chalk mark. Twist a piece of suspension wire through each screw eye so that it hangs down 6 inches below the ceiling line. Cut runners so that they cross the strings exactly at a notch that will accept a cross T (these notches are 3 inches apart along the length of the runner) and hang from the wires so that they just touch the strings. Check that the strings are taut and level as you work. Runners are to be butted, or attached, to span long distances, depending on the product.

9 Setting the Cross Ts. Using the strings as guides, attach the cross Ts between the runners. Along the wall, cut the cross Ts to fit between the inside of the wall angle and the runner. Attach the cross Ts to the runner and sit the other ends on the lower lip of the wall angle.

10 Installing the Panels. When the grid system is complete, simply slide the panels through the grid at a diagonal and let them flop down into position, adjusting them as necessary. At the walls, cut panels to fit with a utility knife. If you are installing illuminating panels (see next page), first bring power cables to the panels from a junction box in the ceiling. Make sure power is off before attaching the cables at the junction box. Get an extra pair of hands to raise the illuminating panel into position in the grid and wire the panel to the power cable. Then set the rest of the panels in place.

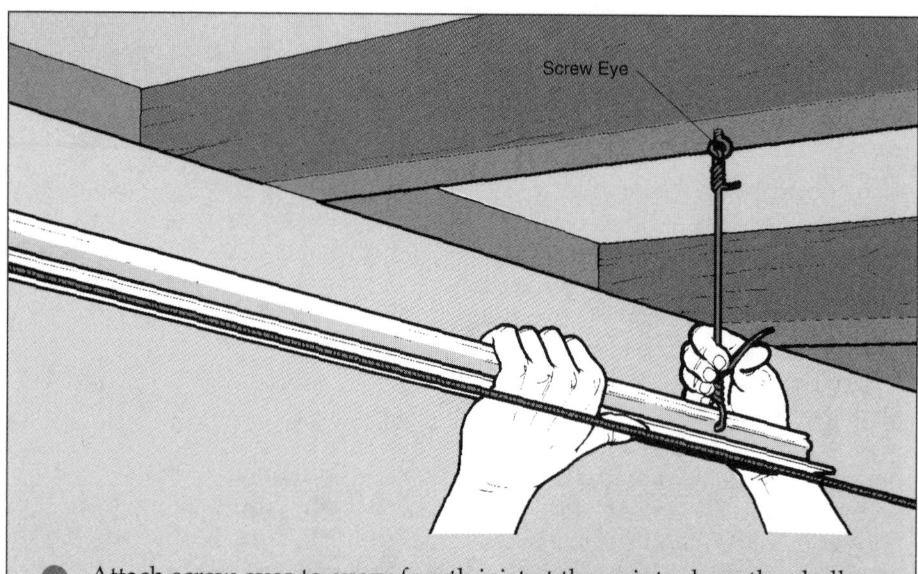

8 Attach screw eyes to every fourth joist at the point where the chalk line crosses it and twist wire through the eyes so that about 6 in. of wire hangs below the height of the wall angle. Attach the runners with the bottom edge at string level.

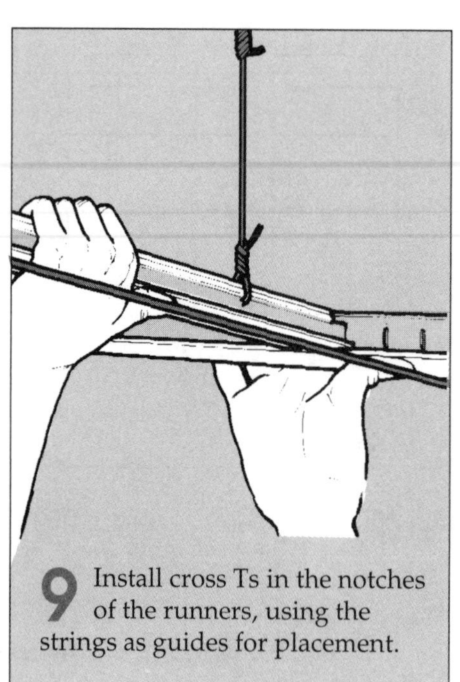

9 Install cross Ts in the notches of the runners, using the strings as guides for placement.

10 Install and wire any lighting panels. Then, angle panels through the grid and set them in place.

Boxing in Pipes

1 Hanging Runners Below the Obstruction. If a pipe or duct intrudes below the level of the ceiling, you can box it in with members of the grid system. You will need U-shaped channel molding and extra wall angle, both twice the length of the box. Figure the box, or break as it is called, into your original layout plans and leave the ceiling open at that point. Attach suspension wires to the ceiling along both edges of the gap and hang runners from them at least 3 inches lower than the pipe. Attach wall angle to hold runners where they meet the walls and fasten the runners to the wall angle at both ends with pop rivets (a pop riveter is an inexpensive tool available at any hardware store). Drill the pieces one at time. Mark the wall angle and drill it, then mark through the holes onto the runners and drill them. Install cross Ts in the slots at the required intervals between the runners.

2 Attaching Channels for Vertical Panels. To keep the side panels in position, pop-rivet U-shaped channel molding to the runner at the edge of the break, and wall angle to the runner hanging directly below. Drill as described above.

Cut panels to fit the sides of the break and install them separated by pieces of cross T cut to fit the channels.

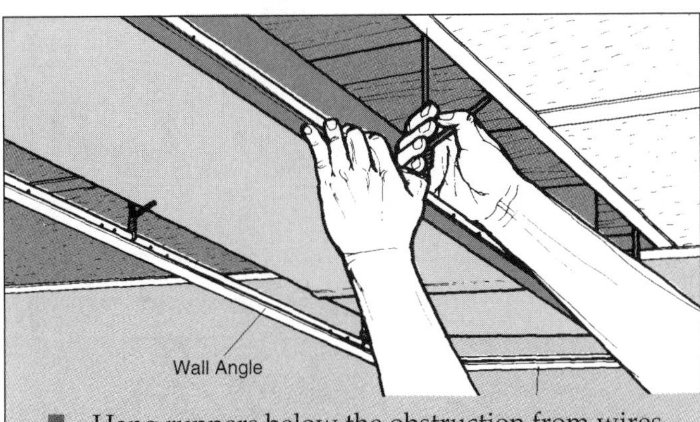

1 Hang runners below the obstruction from wires attached above the edges of the main ceiling; use strings (as above) to mark the level. Attach the runners to wall angle with pop rivets through pre-drilled holes, where they meet the wall.

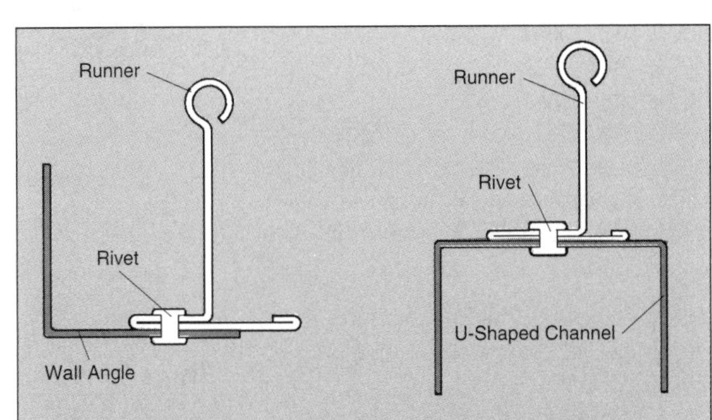

2 Pop-rivet U-shaped channels to the runners, on either side of the gap in the main ceiling, and wall-angle to the runners hanging below in the configuration shown here. Panels cut to fit sit in these holders held in place by cut-down cross Ts.

Installing Illuminated Ceilings

There are several ways to bring light through a suspended ceiling:

- Buy pre-wired luminous panels that sit in the grid the way regular panels do;
- Install fluorescent fixtures on the ceiling in one or more places, with translucent plastic panels in the grid below;
- Run fluorescent tubing over the entire ceiling and suspend a ceiling composed entirely of translucent panels below.

A good-size fluorescent fixture to use in a luminous ceiling is a 4-foot length of 40-watt rapid-start lamp. To determine the number of lamps needed, sketch out the dimensions of the ceiling planning for the lamps to lie in parallel lines between 18- and 24-inches apart. (The narrower spacing gives a more even light, but is more expensive.) On both ends, allow 8 inches between the ends of lines and the wall.

For long runs of fluorescent tubing, connect the fixtures end-to-end with special lock nuts and connectors available where you buy the fixtures.

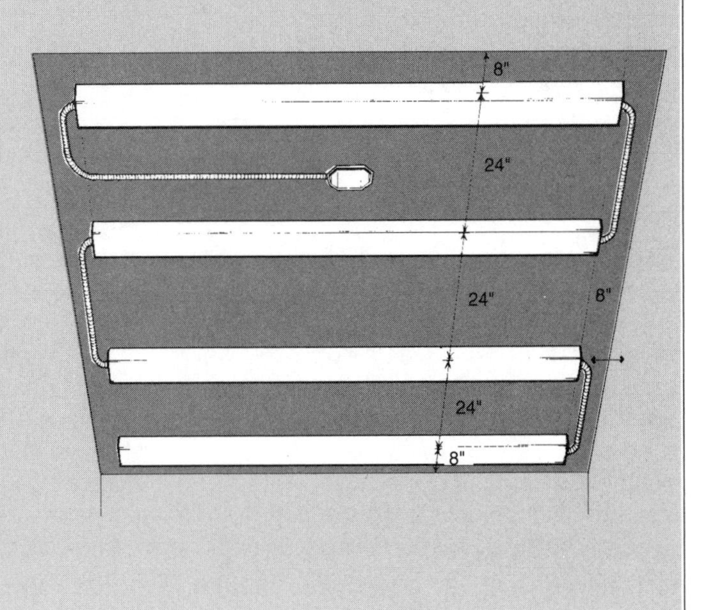

Soundproofing Ceilings

If you are building a new ceiling in a room where soundproofing is a concern, or would like to solve problems of sound control, this section describes alternative solutions. Unfortunately, soundproofing only just part of the structure of the room you want to make quieter will probably have little effect.

Sound passes through ceilings from above and through ceilings under floors from below. Impact noise on a floor above is most common, hard heels on a bare wood floor is the worst offender. For soundproofing purposes, regard a room that has living spaces both above and below as having two ceilings: one that it sits on, and the one above.

The principles of soundproofing a ceiling are the same as those of soundproofing a wall, but the number of possible interventions is more limited because the structure of a ceiling and floor above does not permit doubling the barrier to sound (as in building a double wall with dead air space between the two parts), or building with staggered studs (because the joists that hold up the roof or second story cannot be staggered). The most effective way to soundproof a ceiling is to insert insulation and hang the ceiling surface from Z channels—metal strips that hold wallboard away from an old ceiling surface— or bare joists. They are used the same way resilient channels hold wallboard away from a wall.

The best time to soundproof a ceiling is before the ceiling is finished. With finished ceilings, you may choose to install a new surface over the old, sandwiching insulation between the two, or to rip down the old ceiling and build a new one, perhaps with a double skin. Impact noises can be treated without touching the ceiling at all by padding and carpeting the floor above.

Quieting Above the Ceiling

A multi-layer cushion will eliminate sharp raps above your head and deaden, but not eliminate, the sound of heavy footfalls. First, a layer of 1/2-in. insulating board is nailed to the floor, followed by a layer of hardboard glued to the insulating board. This in turn is topped by a layer of rubber underlayment glued to the hardboard. Wall-to-wall carpeting is then laid on top of that.

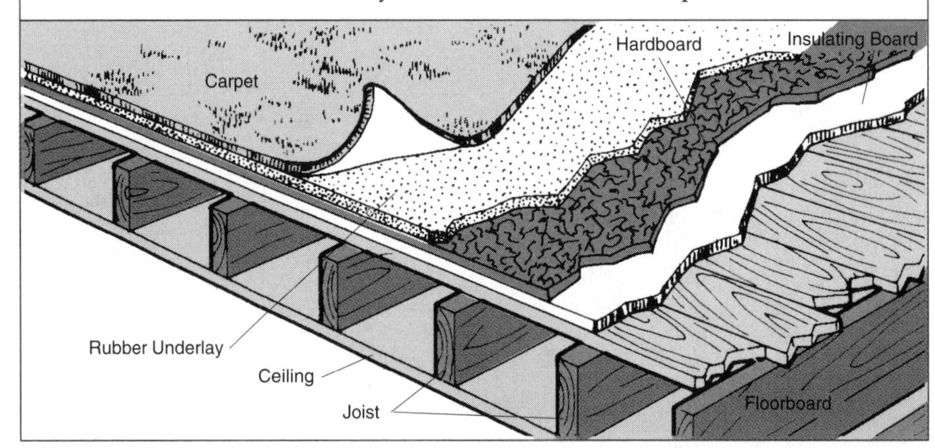

Soundproofing without Breaking Into a Ceiling

1 Marking for Z Channels. Find the direction and spacing of the joists above the ceiling, using the methods shown on page 13. Mark the position of the joists, then snap chalk lines across them (at right angles to the joists) about 6 inches out from either wall.

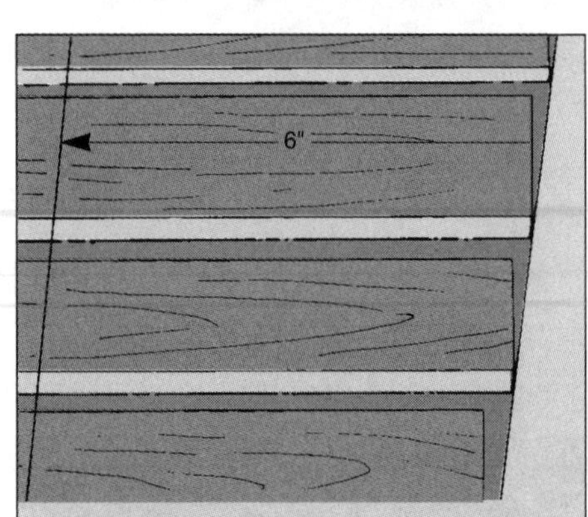

1 Probe to find the direction of the joists, then snap chalk lines across them about 6 in. from both walls.

2 Installing the Z Channels. Divide the distance between the walls into intervals convenient for working with wallboard—in this case 24-inch spacing will be sufficient. If the intervals do not come out to an even 24 inches, plan the installation so that any sheet ending at a wall is supported by at least two Z channels. Mark for these strips so that the middles of the lower flanges fall on the interval marks (to allow for the offset of the flanges from their attachment points). Attach the Z channel across joists with 1½-inch wallboard screws. Orient the strips so that all of the flanges are facing in the same direction; where pieces will not reach from wall to wall, butt overlap the ends and screw through the doubled channel.

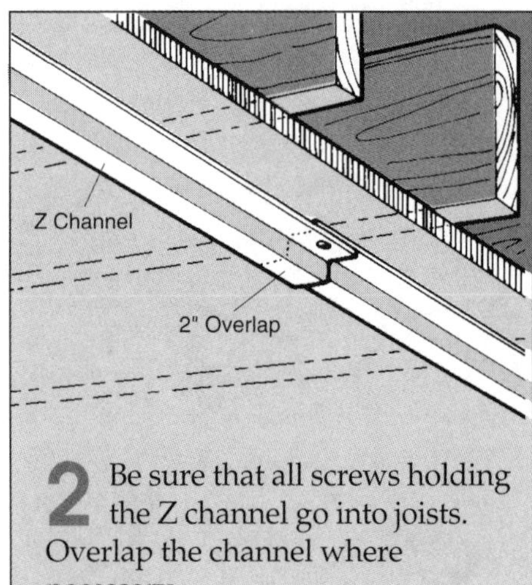

2 Be sure that all screws holding the Z channel go into joists. Overlap the channel where necessary.

3 Installing the New Ceiling. Use the techniques described on pages 56-57 to raise the wallboard and attach it to the Z channel with 1-inch wallboard screws driven through the metal flange. Leave a 1/8-inch gap along the walls for caulking. Where pieces butt, center the joint on the Z channel and attach the second piece to the channel before the first, in order to put the least strain on the flanges. Fill the gap around the wall with acoustic compound.

4 Insulating the Ceiling. After completing one row of wallboard, slide 2-inch insulation between the wallboard and ceiling above. Continue to install the wallboard in this fashion. At the last strip, glue the insulation to the wallboard between the lines where it will meet the channel.

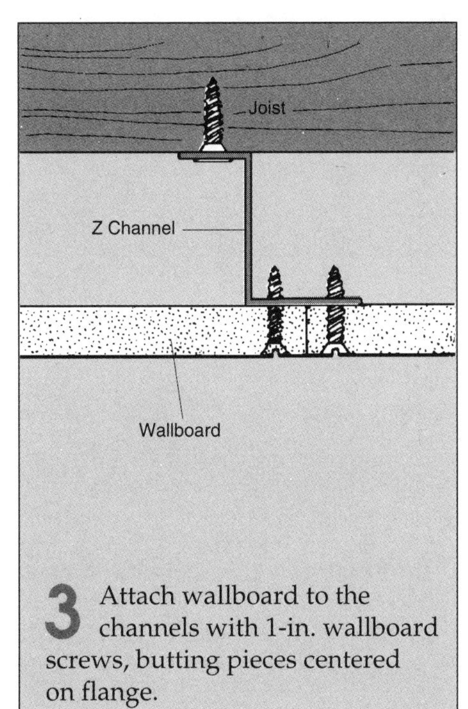

3 Attach wallboard to the channels with 1-in. wallboard screws, butting pieces centered on flange.

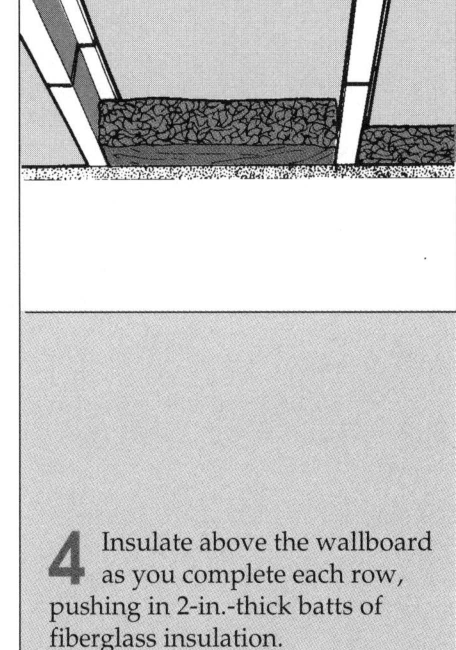

4 Insulate above the wallboard as you complete each row, pushing in 2-in.-thick batts of fiberglass insulation.

Installing a New Ceiling

1. Removing a ceiling to replace it with wallboard is relatively easy if the existing ceiling is wallboard. Simply cut into the wallboard (checking first for electrical wiring and other hazards as described on page 14), and pull it down. Be careful not to pull pieces down on your head. Wear protective goggles.

2. Fill between the exposed joists with 6-inch insulation, barrier side down and staple it to either side of joists.

3. Attach Z channels perpendicular to the joists, using the same procedure as for covering a finished ceiling (described above). Follow by covering with wallboard.

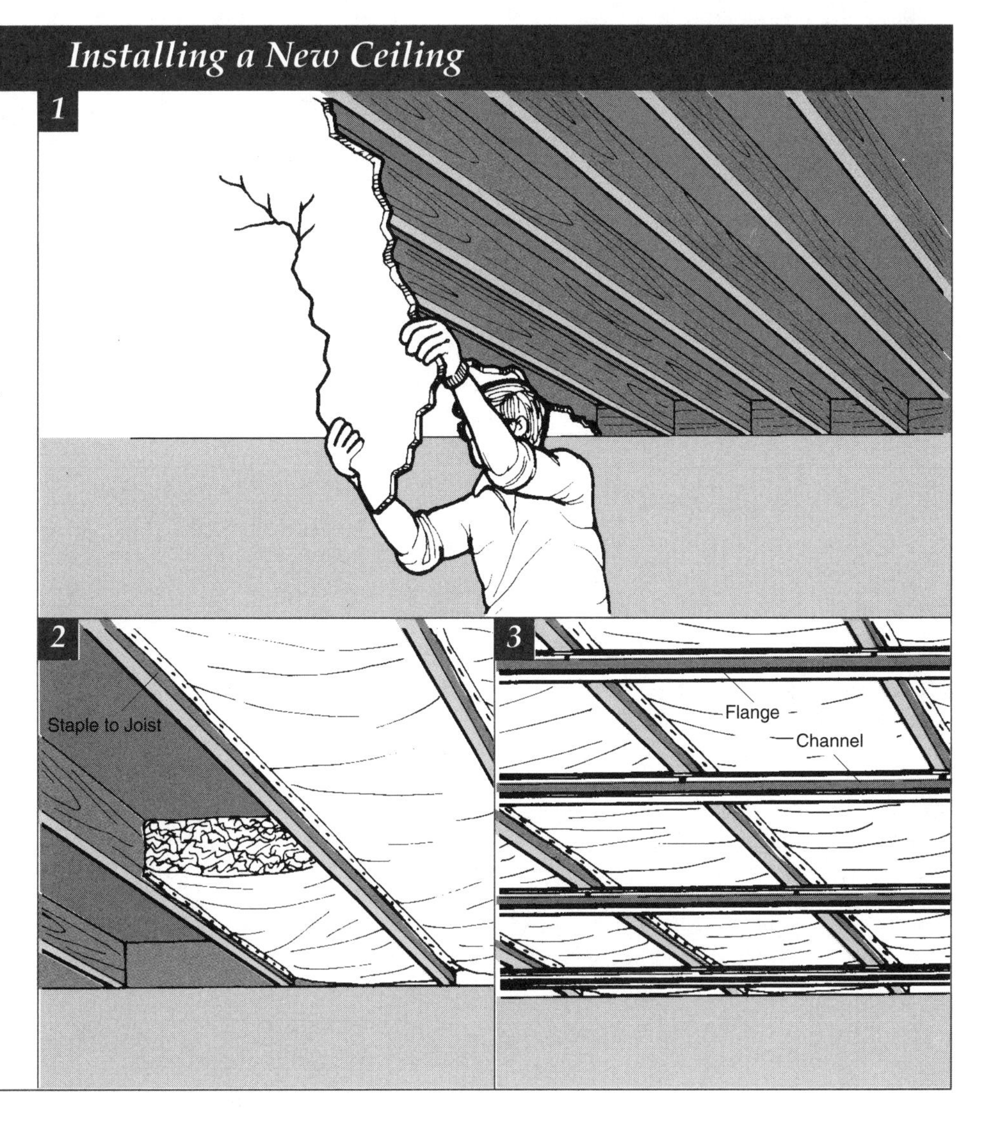

Finishing Attic Ceilings

Finishing an attic is a relatively inexpensive way to add living space to a home. An attic offers attractions many basements cannot—including access to natural light and natural ventilation. The walls and floors in an attic are essentially the same as in the rooms below (see page 22) but ceilings are somewhat different. Before anything else, you must check the usable headroom in an attic. A ceiling must be a minimum of 7 feet high.

Special measures must be taken when building an attic ceiling. Because all or some of a ceiling is attached to the inside of the roof, it must have provisions for stopping heat in all but perfect climates. If you do not insulate, you may wind up with an attic office or bedroom that is stifling on summer days and hard to keep warm in the winter.

There are two basic variations in attic ceiling treatments: a flat ceiling or a cathedral ceiling. The flat ceiling is built on collar beams that serve the same function in the ceiling as joists do in the rooms below. It follows the roof line to a knee wall and this creates odd angles to deal with where the ceiling meets the roof and the wall (see opposite page). You can use strings hung from nails to simulate such a ceiling to get a sense of what the space will look like, as you plan the conversion. The ceiling should be insulated along the roof and between the collar beams.

A cathedral ceiling gives a sense of greater space and eliminates the need for ceiling framing. It also looks good with sheet or plank paneling that goes on the same way as on walls, but with special provisions for insulation.

If your attic already has collar beams supporting the roof, you can still leave the ceiling open to the top of the roof. The beams can be used to support a storage platform.

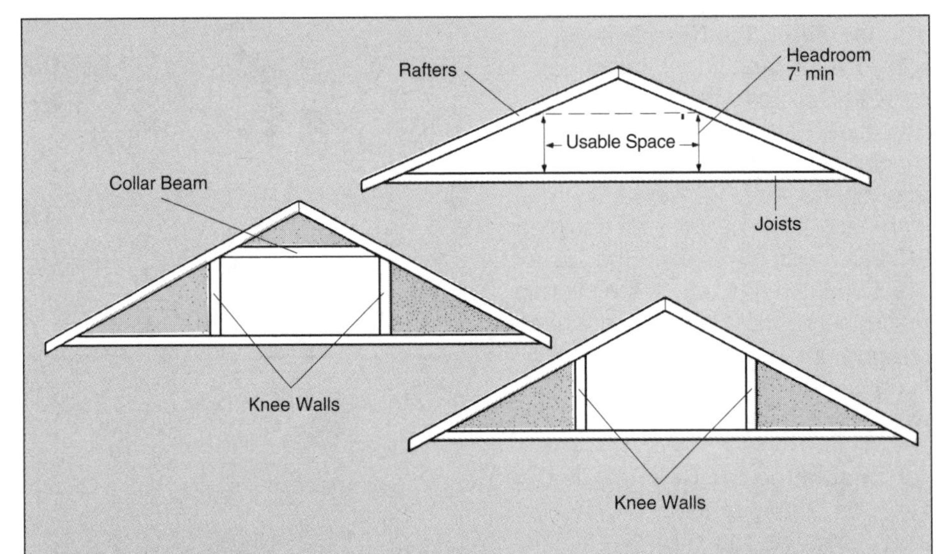

Planning the Conversion. Measure the amount of floor space in an attic that falls under portions of the roof that are at least 7 ft. high—this is the amount of livable space you have to work with. You can set the knee walls farther under the roof and use the lower space for storage or sleeping.

Building a Flat Ceiling

1 Installing Collar Beams. Measure the height of the ceiling and mark it on the inside edge of opposing rafters, then measure between the outside edges of these rafters at the ceiling height to obtain the collar beam length. Cut 2x6s to that length less 1/2-inch, then measure the angle of the roof slope and cut the ends of each beam to fit. Attach the beams to the same side of each air of rafters by driving two 16d nails through the beam into the rafter, then two from the other side of the rafter. Toenail blocking between the rafters to provide nailing surfaces for the flat ceiling and for the part of the ceiling that follows the roof line.

2 Insulating the Attic. Check the insulation requirements for your area and buy batt or blanket insulation with the appropriate R-value and appropriate width. Fill the part of the floor under the eaves that will be walled off, taking care not to block the ventilation that is necessary to keep the attic dry. Staple blankets (vapor barrier facing in) or stuff batts between studs, rafters and collar beams, making a snug fit where the ends of batts or blankets meet.

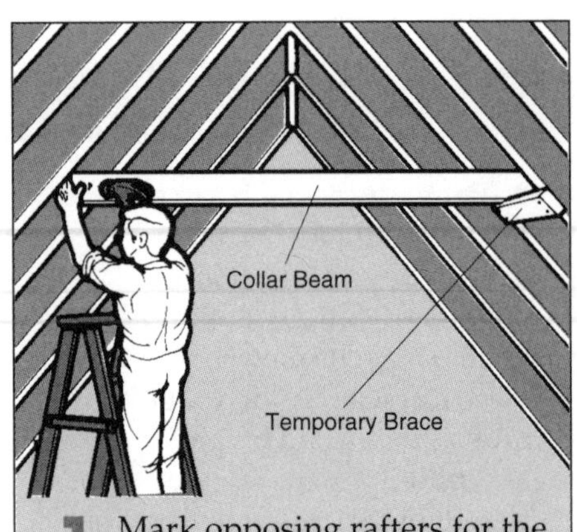

1 Mark opposing rafters for the height of the ceiling, measure the angle, and cut collar beams to fit.

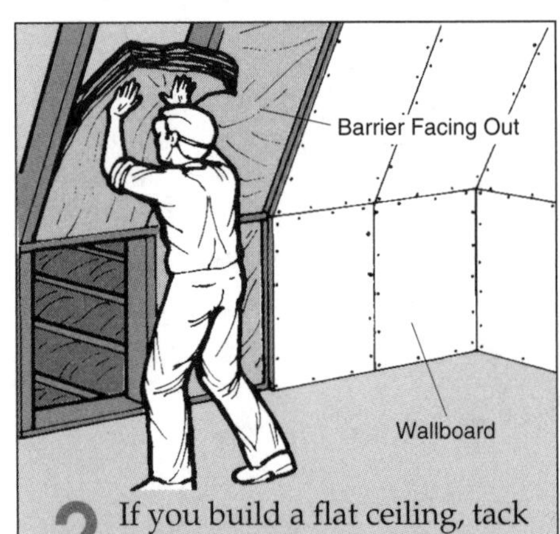

2 If you build a flat ceiling, tack insulating batts between the beams before putting up wallboard.

Plastering Obtuse Angles

1 Covering the Seam. If you have a reasonably steady hand you can hide the seams where a flat wall meets the roof line by sculpting it with patching compound (see pages 30-32). Spread on a fairly thick first coat with strokes perpendicular to the seam. Try to keep as regular a line as possible, feathering the edges and smoothing ridges between strokes. Let the first application dry and then smooth it with sandpaper and a damp sponge.

2 Finishing the Seam. Fill the angle with a second layer of compound, feathering the edges and smoothing the angle further. Allow to dry and then sand. Finish with a few more coats.

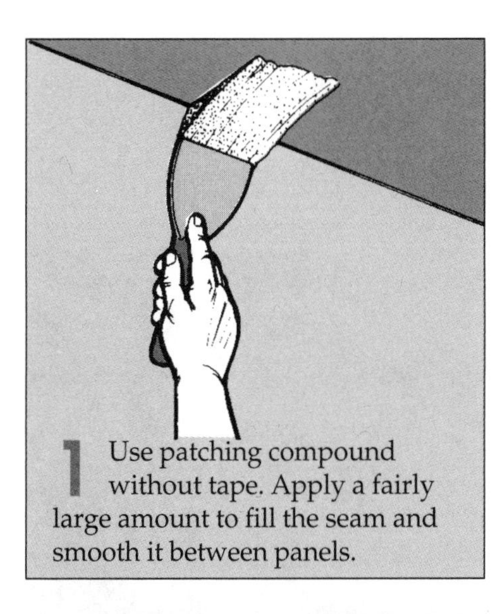

1 Use patching compound without tape. Apply a fairly large amount to fill the seam and smooth it between panels.

2 Let the first application dry, then apply additional layers, smoothing and sanding until the seam is regular.

Wallboard Cathedral Ceilings

If you are finishing an attic, insulate between the rafters as described on page 66 and cover with wallboard horizontally. Tape the seams and tape the peak of the ceiling as you would an inside corner. You can install sheet paneling over the wallboard.

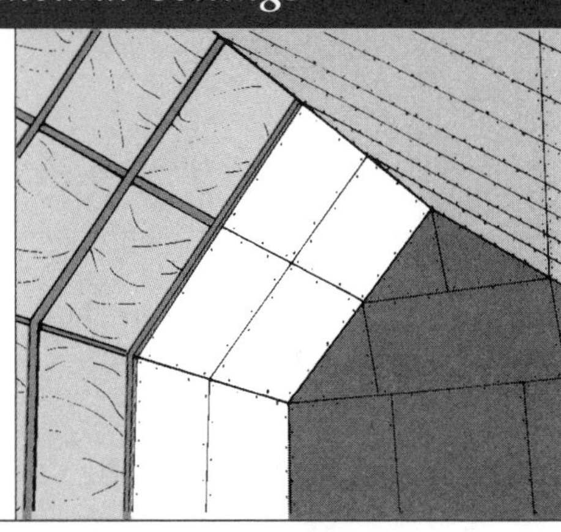

Paneled Cathedral Ceilings

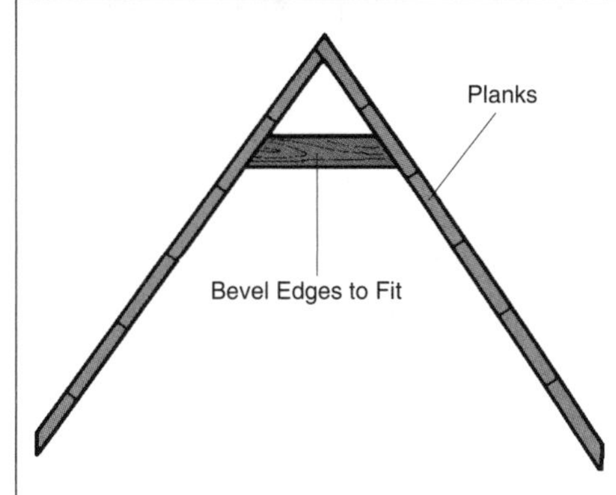

If you want to finish a ceiling with plank panels, especially effective in attic conversions, nail planks across rafters, fitting them as you would on a wall. Panel the ceiling to within a few inches of the peak and measure the angle at which the sides of the ceiling meet and bevel a plank to fit between the converging surfaces.

Cathedral Ceiling with Exposed Collar Beams

If the structure of your house requires the retention of existing collar beams for roof support, fit the ceiling surface around the beams. Collar beams below a cathedral ceiling can provide support for a sleeping loft, if the roof is high enough, or for storage. For storage, use 1/2-inch plywood for flooring between beams.

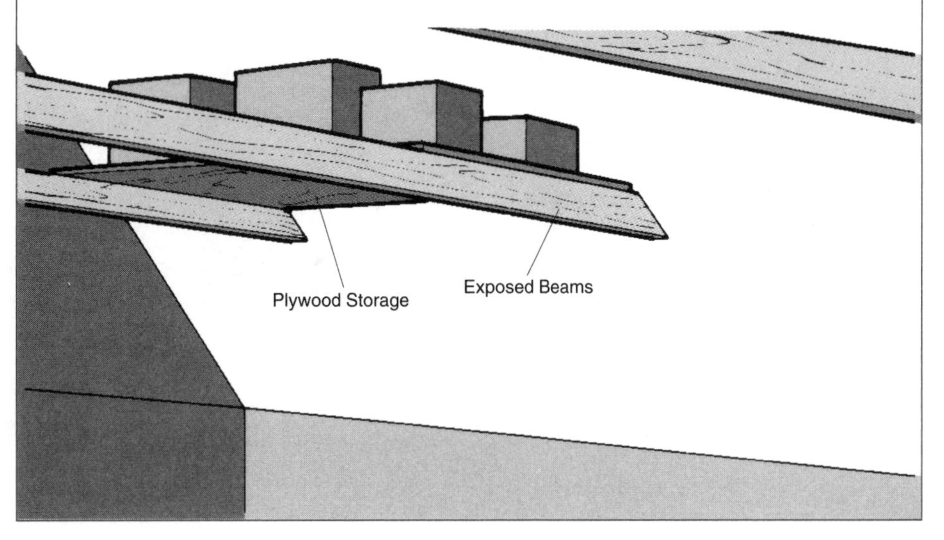

Installing Skylights

Bringing daylight in all its variations through a ceiling is an attractive way to light a room. A skylight can serve as a window to look through or as a light source at the top of a well that opens into the ceiling a few feet below. You can buy or build skylights that open for ventilation or units that are sealed. The job of installing a skylight is somewhat more complicated than other projects in this book. Whether you do it yourself or have the job done for you, read all the directions below to inform yourself as to the planning and purchasing of the appropriate unit.

There are a number of different kinds of skylight windows and ways to install them. They can be constructed from scratch and glazed at home, purchased as plastic domes (or other kinds of acrylic and glass fittings) to be mounted on a curb you build and seal yourself, or purchased as prefabricated units that are installed directly on or through the roof. This section gives instructions for installing a purchased skylight on a curb. The information here also can be used together with the manufacturer's instructions for installing a prefabricated unit. Glazing your own skylight is outside the scope of this book; ask your glass dealer for the best sources of information.

Whether you build your own or buy a ready-made skylight, you will have to cut through the roof and the rafters below and frame around the opening, to keep the roof structurally sound. If there is a crawl space or an attic between the roof and the ceiling below, you must open the ceiling and frame and finish a light well up to the roof. In most cases, a skylight is set on the roof if it sits over a light well. Where it opens through the roof directly into a room, it may sit in the opening or not, depending on the kind of unit. Prefabricated skylights usually come with instructions for specific models and dimensions that make planning easier.

To install a skylight you must work on the roof and in attic or crawl spaces below which can be awkward (it is less difficult to work with a flat roof or cathedral ceiling). Use ladder brackets (they hold a ladder along a roof from the peak) or toe board jacks (they hold planks along the roof to ensure good footing). The nature of the space under the roof will dictate whether you should cut through the roof and work your way down, or cut up through the ceiling first. Do not attempt the project if you are not comfortable working on the roof.

When you plan the location of a skylight, try to determine the structure between the ceiling and roof at that point—you may want to change your plans to avoid a complicated framing problem. Also, put a piece of cardboard on the roof where the skylight will lie to check whether it looks alright from the outside. Check again later, with the actual skylight in place, but not attached, to be sure it looks straight.

Framing the walls of a light well is much like framing walls for rooms, except that the work may involve surfaces meeting at angles other than 90 degrees. Careful measuring and cutting of angles on framing members is the only extra work required.

Besides the skylight itself, you will need lumber to frame openings, finishing materials for a light well (if there is one), and roofing cement, metal flashing, and perhaps some extra shingles, for sealing the skylight at the roof. The work can be accomplished with ordinary carpentry tools: a circular saw (a reciprocating saw will make the job go faster), a stout chisel, and measuring tools including a level, plumb bob, and a sliding T-bevel (an adjustable straightedge for measuring angles).

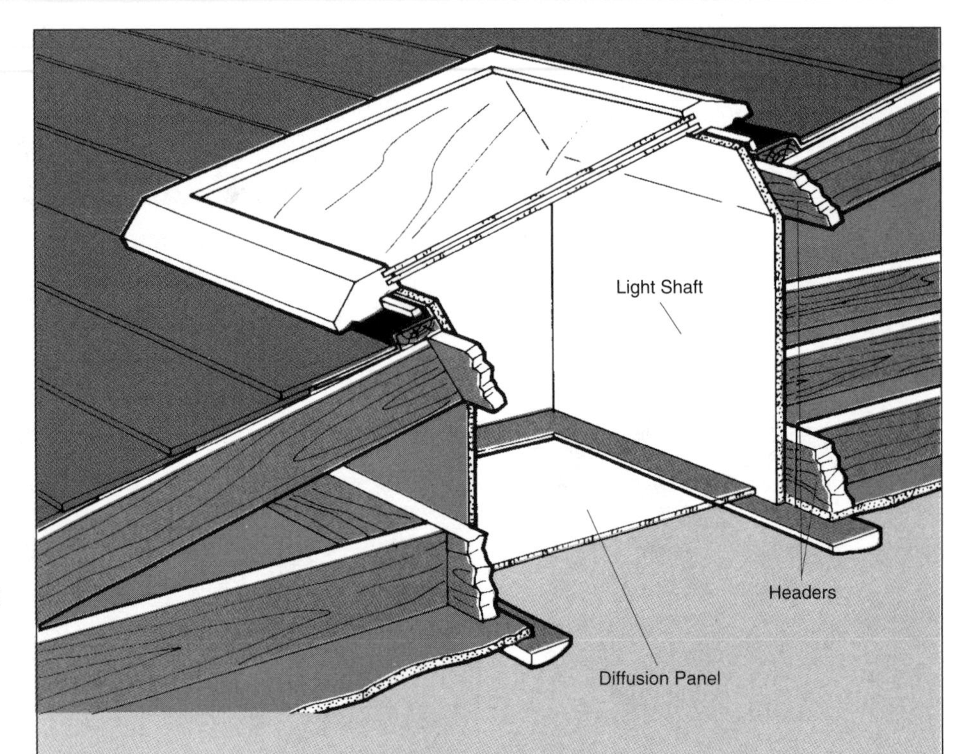

Anatomy of a Skylight. A skylight opening through a roof above an attic lets light into the room below through a light well. Installing the skylight and constructing the light well require cutting through the roof and ceiling below and framing and covering short walls between the openings.

Building the Curb

Curb-mounted skylights require the construction of a simple open box of lumber called a curb. The purpose of the curb is to elevate the skylight above the roof to permit sealing the roof around it. Use 2x6 lumber, unless otherwise specified in the manufacturer's directions. Measure and cut the lumber, making sure it is not warped, so that the box will have the inside dimensions specified for the skylight. Pre-drill for nails at the butt joints and attach the pieces.

To be sure that the curb is square, measure between opposite corners (diagonally across the curb); the two diagonals should be the same length. Keep the curb from getting out of square by tacking pieces of strapping across opposite corners.

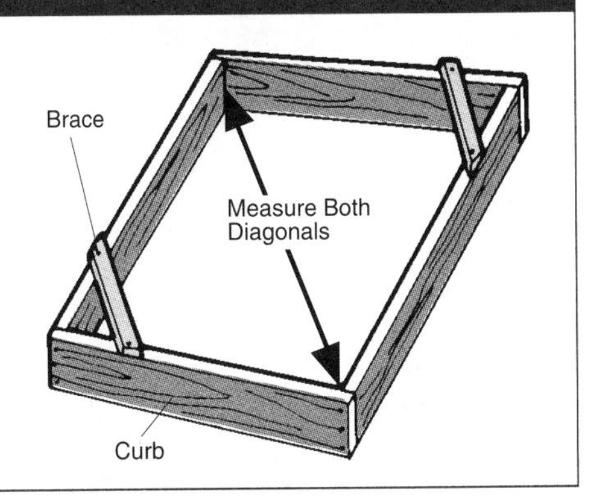

Adding Flashing

Flashing consists of pieces of aluminum bent to fit against the curb and the roof to keep out water. You can bend flashing to fit all four sides of the curb or only the top and bottom, with the sides covered by step flashing (see page 71). The aluminum can be purchased from hardware stores or lumberyards and cut to fit with shears. Bend continuous flashing so that the piece at the bottom will fit onto the curb first, the sides will overlap the bottom piece, and the top will overlap the side pieces.

To facilitate the bending process, make full-size paper templates to be sure your bends will work; then use a straight 2x4 for the actual bending—working against the curb itself can be awkward. The flashing will be attached to the curb when the curb is mounted on the roof.

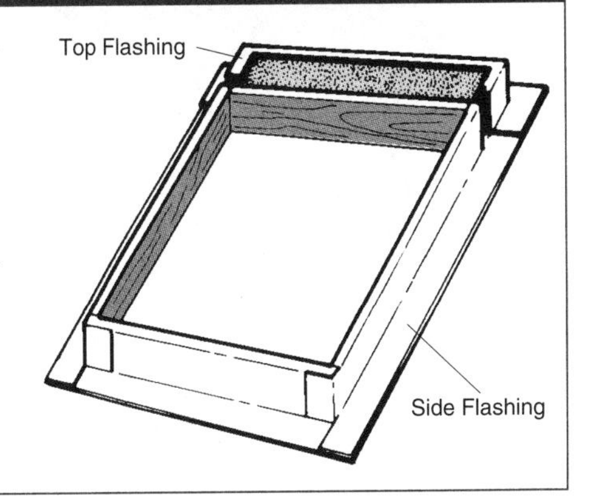

Siting the Skylight

If you can work above the ceiling, mark the center of the place where you want the skylight and drill a locater hole. Make adjustments to minimize framing around openings. Try to butt at least one side of the skylight to a rafter (left). Outline the area to be cut from the roof (the inside dimensions of the curb) and, for a straight light well, locate the ceiling cut directly below with the plumb bob (right). Mark through the roof with a drill at the corners of the opening. For other light wells (see page 72), use a T-bevel and strings to mark off the angles and the locations of the corners, run strings from the corners of the roof opening to the floor. If you can't work above the ceiling, probe through it to find joist locations and cut a section of the ceiling away between joists. Measure and mark the cut in the roof, working through the hole in the ceiling. If this is not practical, cut an entry hole in the roof.

Header Plans

Rafters that are cut to make the opening for the skylight must be secured with headers. If the skylight does not fit between existing rafters, jack rafters must be installed to create framing of the required width. Double the headers when bridging more than two rafters.

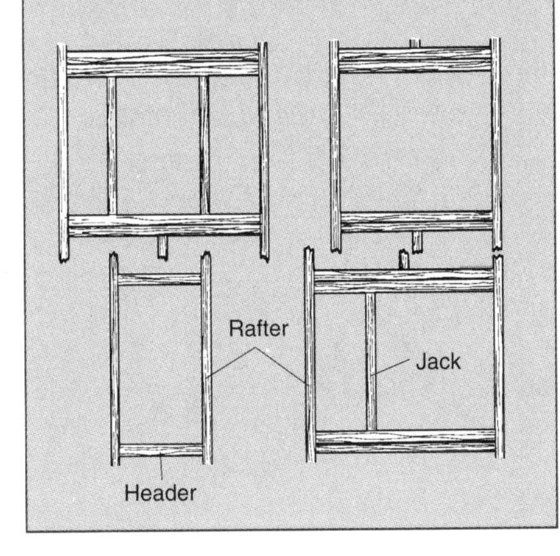

Installing the Skylight

1 Cutting the Roof. Cutting the typical roof covered with shingles or shakes is a two-step process: First the roof covering is cut, then the roof sheathing beneath. The curb sits on the roof sheathing, supported by rafters and headers beneath.

2 Marking the Roof. The inside dimension of the curb is the dimension of the actual hole through the roof; the material that covers the roof is cut back to the outside dimension of the curb. Put the curb in position on the four nails that indicate its proper location and mark its outside outline on the roof with chalk.

3 Cutting Roofing Material. Use a circular saw with a combination blade to cut wood shingles, a chisel or utility knife to cut asphalt, or shears to cut a tin roof. If using a circular saw, set the blade depth to the thickness of the shingles so as not to cut the sheathing beneath. Cut around the marked outline and pry up the roofing material. Save asphalt shingles for any patching necessary around the curb.

4 Cutting the Sheathing. Snap chalk lines between the nails marking the inside dimensions of the curb and cut the sheathing with a circular saw set to the depth of the sheathing. Pry the piece of sheathing off the rafters.

5 Framing the Opening. Measure 1½ inches in from the hole (3 inches for double headers) along the rafter or rafters that cross the hole you have cut in the roof; mark them at right angles to the roof line with a square. Support the rafter you will cut, with props of 2x4s nailed to the side of the rafter on either side of the section to be removed, and make the cuts. The top flashing, and then the curb, are installed before the headers so that the curb can be attached from below (through the sheathing where the headers will sit).

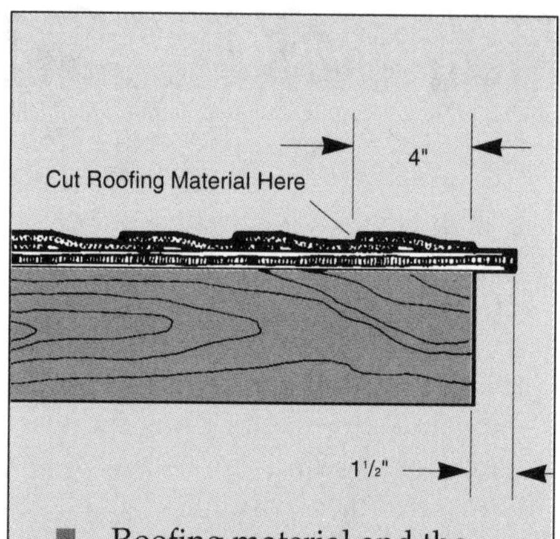

1 Roofing material and the rafters below are cut back to leave 1½ in. of roof sheathing for mounting the curb and for attaching headers.

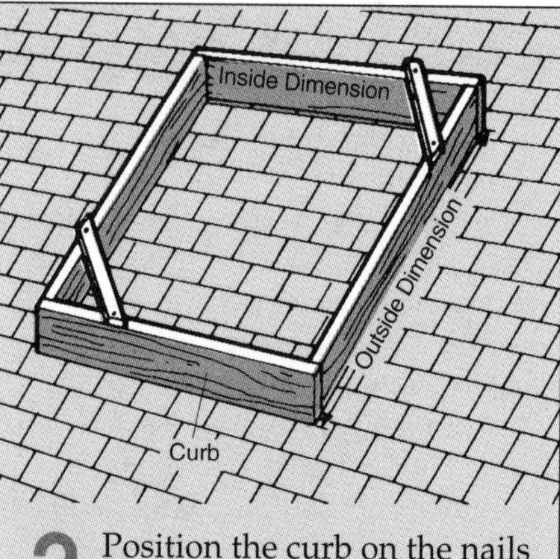

2 Position the curb on the nails marking the location for the skylight and mark the inside and outside dimensions.

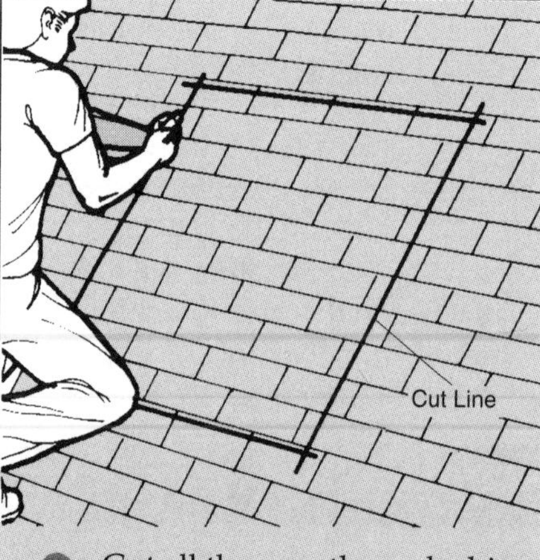

3 Cut all the way through shingles so pieces outside the outline do not pull up with those inside it.

4 Pry out nails along the line of the cut so the circular saw blade will not be damaged when cutting.

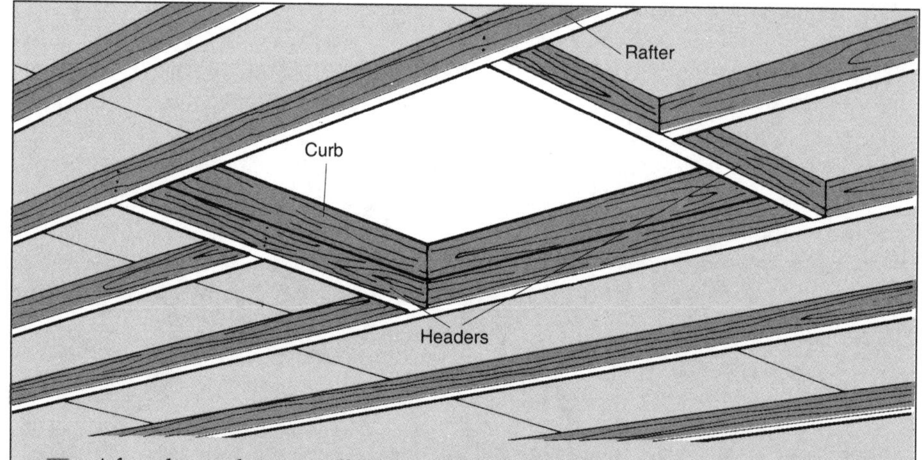

5 After the curb is installed, support for the interrupted rafters is added below. The headers should be made of lumber which is the same dimension as the rafters, and attached with nails through rafters into header ends and into the ends of the cut rafters.

6 **Installing the Top Flashing.** Pull the roofing material back about 4 inches from the hole in the roof, cover the sheathing beneath with roofing cement—spread it with a spatula—and slip the top flashing under it. Cover the flashing with a layer of roofing cement and lay the roofing material back down. Be generous with the roofing cement. Too much is no problem—too little is a problem.

7 **Seating the Curb.** Set the curb in place. Use 12d nails to attach the curb from below, nailing up through the sheathing. Where the curb sits atop rafters, toenail down through the curb into the rafters. If you are installing jack rafters, nail up through the sheathing where the jack rafters will sit. When the curb is secured, finish the framing between rafters below.

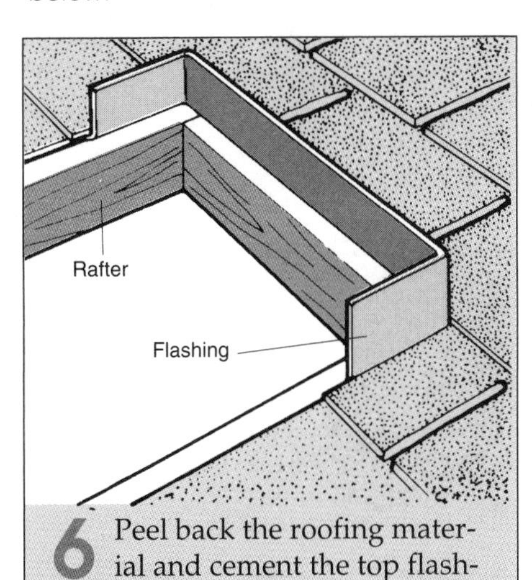

6 Peel back the roofing material and cement the top flashing in place for the curb to fit into.

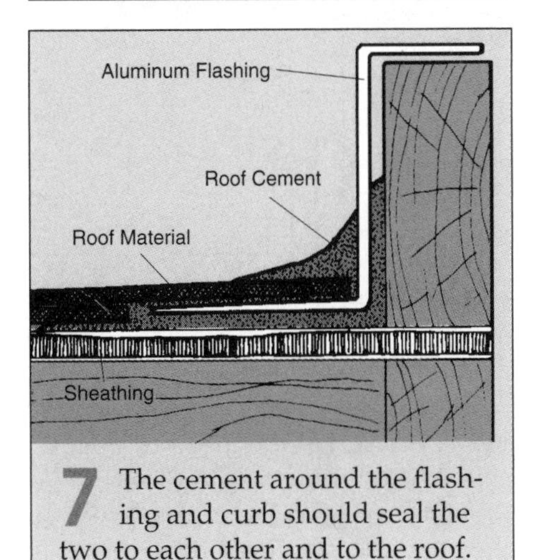

7 The cement around the flashing and curb should seal the two to each other and to the roof.

8 **Adding Step Flashing.** Attach the side flashing pieces, then the bottom piece, as described above. Nail the flashing to the top of the curb with 1-inch galvanized nails coated with caulking compound. Nail the roofing material back down on the roof outside the edge of the flashing, with roofing nails dipped in roofing cement, and cover the nail heads with more cement. You also can use small pieces of flashing material overlapped along the sides.

9 **Seating the Skylight.** Apply sealant to the top of the curb and put the skylight in position on the curb. Attach it through the holes around the edge of the skylight with the nails or screws specified by the manufacturer.

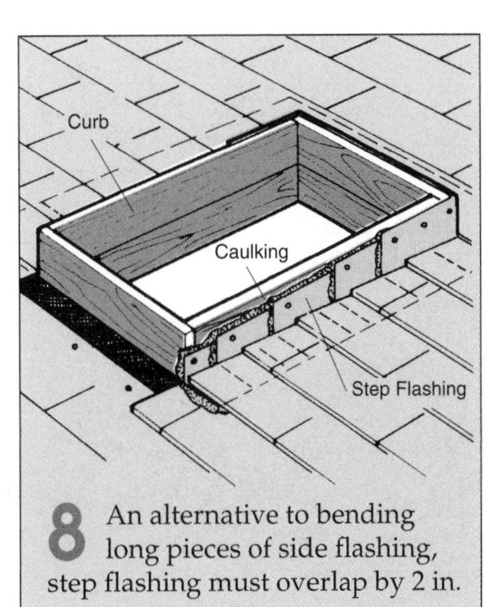

8 An alternative to bending long pieces of side flashing, step flashing must overlap by 2 in.

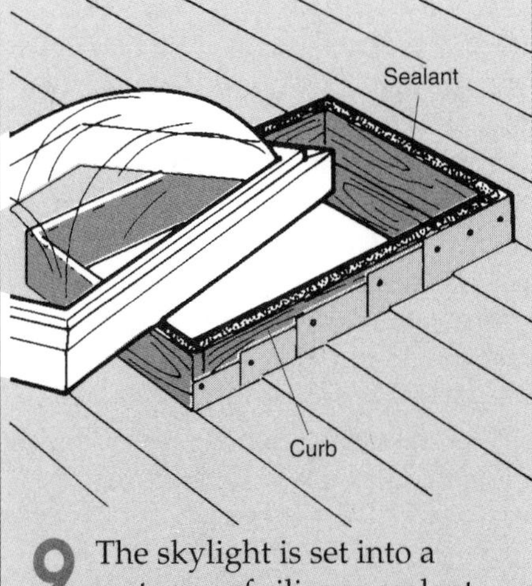

9 The skylight is set into a waterproof silicone sealant spread in a wavy bead along the top of the curb.

Setting a Self-Flashing Skylight

Installation of a self-flashing skylight with a built-in curb is the same as for a skylight curb. Roofing material is cut back to the outside dimension of the skylight, then roofing cement is spread around the hole in the roof and the skylight set into it and nailed down. Roofing cement is applied over the flashing and shingles nailed over it.

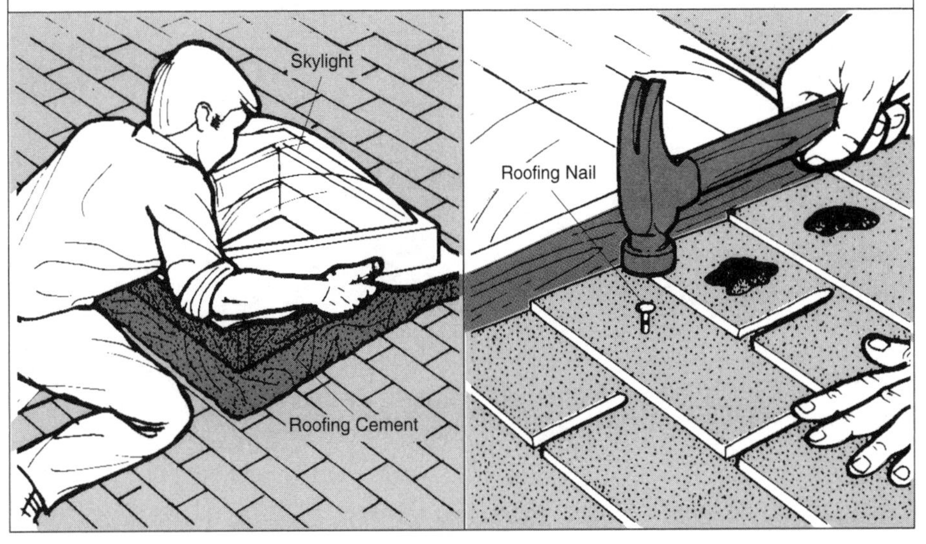

Framing the Light Well

1 Types of Light Wells. Building the walls for a light well is a straightforward framing job but the angles involved are usually different from those in ordinary wall construction. The easiest situation is a light well descending straight down from a flat roof—all of the angles are right angles. In the other situations some studs are cut to fit the slope of the roof line or the flare of the light well. Cut away the ceiling surface at the points you marked (see page 69) to expose the joists above. Tack lengths of 2x4 across several joists, about a foot back from each side of the cuts you must make in them, to keep them from bowing down. Measure back from the edge of the hole in the ceiling a distance equal to the width of the headers you will install on either side of the hole. Use lumber the same dimension as the joists for the headers. If the light well is flared, cut the joists at the angle of the flare; stretch strings from the corners at the top of the well to mark these angles. Frame in the hole in the ceiling as you did the hole in the roof.

2 Attaching Straight-End Walls. Sketch plans for the walls of the light well. They have top and bottom plates as in an ordinary stud wall. In the example shown here, the walls perpendicular to the roof joists are constructed square and are secured to the roof joists by nailing through a wedge as long as the top plate made by ripping a 2x4 to fit the angled gap. Construct these end walls first, then nail them in place.

3 Completing the Light Well. The side walls shown here are built with studs angled at the top to conform to the angle of the roof line. The angle at the top of the studs is the same for all and it is most easily cut in an adjustable miter box. Set the box for the correct angle (double check it) and cut all the pieces at the same time. Measure the height of each stud carefully for a tight fit. Corners are built as for any stud wall.

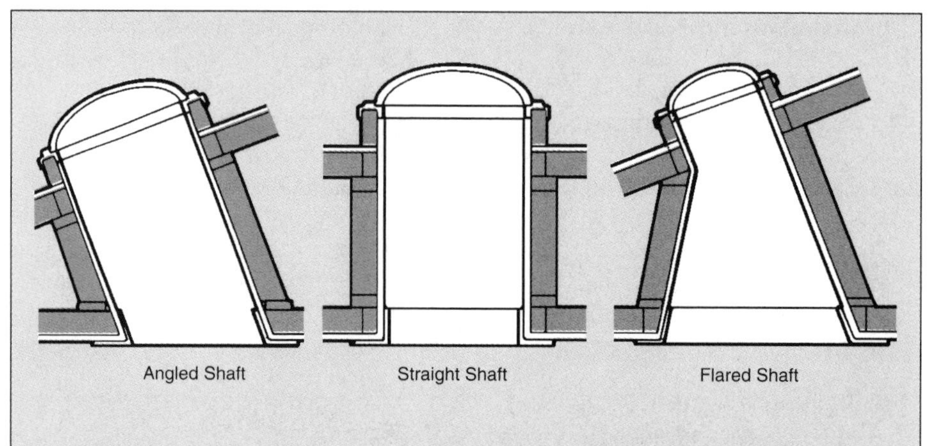

1 Straight-walled light wells are the easiest to construct; angled walls are demanding and require skill. If you want a flared well, you may wish to work with a professional, doing the skylight installation yourself and leaving the difficult framing job to the expert.

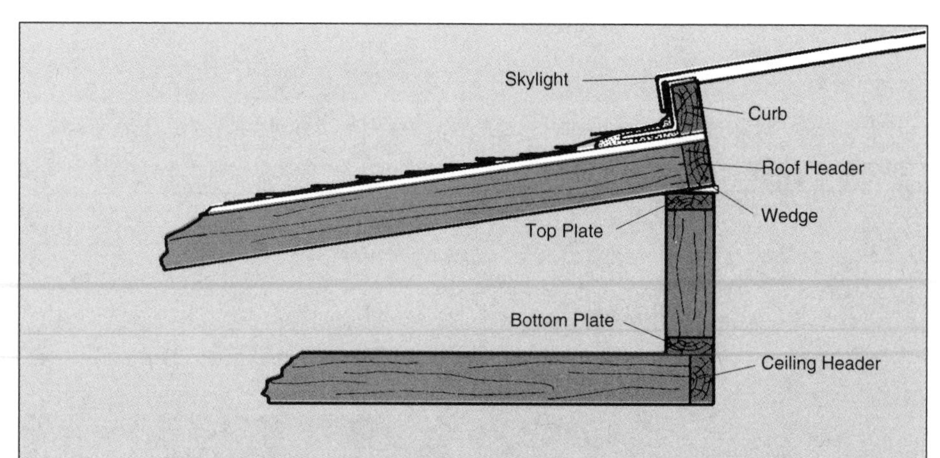

2 A straight-end wall is simply a short stud wall with a wedge between the top plate and the rafters. To cut the wedge, measure the angle of the roof line and set the blade of a table saw, or the shoe of a circular saw, to that angle and rip a piece of 2x4. Tack this to the top plate.

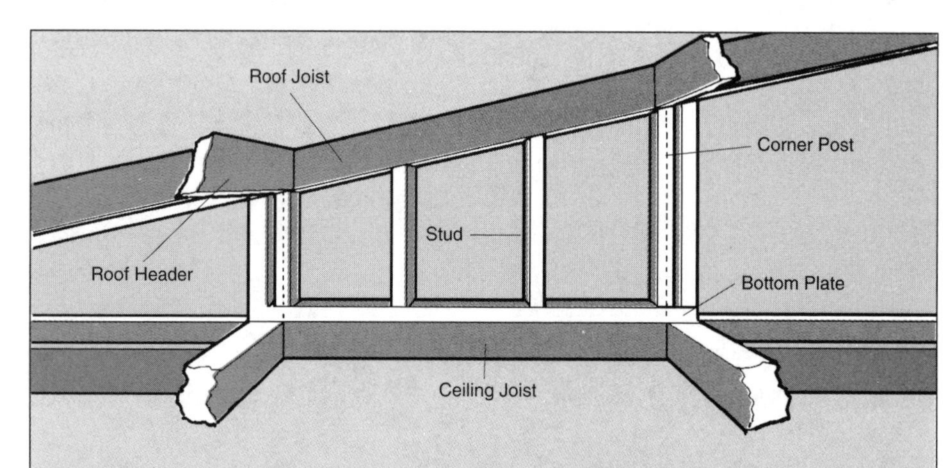

3 Finish the inside of the light well with wallboard as described on pages 26-33. Trim the opening with molding. If you wish to add a plastic diffusion panel at the bottom of the well, set the molding 1/2 in. inside the well as a support or use hardware specially designed for the panel.

Making & Installing False Beams

You can change the appearance of a room completely by adding simulated beams across a plain ceiling. If you have added a structural beam when removing a bearing wall to make a room larger, you can disguise its presence by flanking it with simulated beams.

The prefabricated beams may be purchased or you can construct them out of wood at home.

Prefabricated false beams are molded in plastic and attach to the ceiling with adhesive. Wooden false beams are milled into a U-shaped track with gouges on the outside that look like hand adze marks.

Anyone comfortable with a hammer and saw can fashion false beams with a professional look. Wooden false beams can be shaped and stained or painted for different visual effects. They can be scarred and stained for a rough-hewn look, or sanded and painted, or sanded to soften the edges and stained to resemble a time-worn piece of lumber. For those expert woodworkers, a more challenging job would be adding a grid of beams that cross at right angles to create an elegant high ceiling.

Decorative wood beams are attached to tracks mounted on the ceiling surface. Installing wooden false beams requires locating the joists in the ceiling, but molded beams can be attached to wallboard or plaster between joists.

Building & Installing False Beams

1. Measure the ceiling and decide on a plan, spacing the beams evenly. Mark guidelines on facing walls. If you are installing milled wood beams, go to the next step. If you are installing molded beams, run a wavy line of adhesive along the ceiling and set the beams in place—they are light and easy to handle. If you are building beams, use 1x4s or 1x6s to make the beam itself, gluing and nailing the sides flush with the bottom, either butted or mitered. Set the nails.

2. A beam may be finished with paint, stain, or clear finish. Smooth the corners with sandpaper, a plane, or a rasp, depending on the effect you want. If you are painting, fill cracks and nail holes with wood putty first. If you are staining, fill with putty afterward.

3. Find the joists and use 16d nails to attach lengths of 2x4 or 2x6 (the same width as the false beam) on the ceiling as a track across the joists, or along the joists, depending on your plan.

4. Slip the milled or homemade beam onto the track and attach with nails through the side of the false beam. Set and fill these nails and touch up as necessary.

Installing Ceiling Light Fixtures

Interior artificial lighting is categorized as either general, which illuminates a room as a whole, or task, which lights specific areas. Much general lighting in homes is supplied by ceiling fixtures, as is some task lighting that may be directed at dining room tables or pictures on a wall. Ceiling lights are hung on the ceiling, recessed into it, or hung above it in the case of a suspended ceiling (see page 60).

With the great diversity of bulbs and fixtures available for different purposes—from casting a diffuse light to spotlighting with a focused beam—the range of possible lighting schemes is endless. On this page there are four fairly simple projects that offer dramatic lighting effects. The next two pages give directions for installing fixtures in ceilings under the various circumstances you are likely to encounter in the average home, followed by directions for running power cable to these installations. Finally there are directions for installing track lighting and ceiling fans.

Adding ceiling lights usually involves the two operations of bringing power to the location of the light and installing the fixture. In most situations you must break into existing ceilings, then patch the area after the installation is done. Unfinished ceilings are the easiest to work with; lighting should be carefully planned to take advantage of this before finishing basement or garage ceilings. Ceilings below crawl spaces and unfinished attics are easier than those below finished rooms because joists are exposed. If the attic has rough flooring, it is still easier to open the floor and work from above than to open the ceiling from below. Ceilings under finished rooms must be cut open to give access to a joist, or two joists, depending on the type of ceiling. Running wire to the fixture also may require cutting the ceiling and the wall in a few places to reach a power source.

Valances

A valance is a baffle attached to a wall, cabinet, or ceiling to deflect and concentrate light upward or downward and to conceal the light source, which is usually a fluorescent tube. When it directs the light down, it accents the wall and furniture below; when used to deflect light upward, it bounces light off the ceiling for a soft, general lighting effect. A valance can be attached along a wall with nothing to stop the light shining out from the top or bottom to combine these effects. The best results are obtained using distances in the installation that approximate those given in the drawing here.

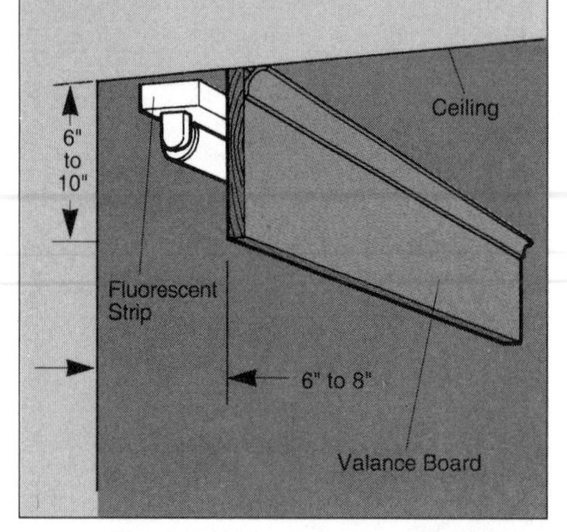

Coves

A cove baffle to direct light creates indirect, general lighting. The base of the cove is installed not less than 12 inches from the ceiling, parallel to the ceiling and perpendicular to the wall, with the lighting fixture attached to the wall just above the cove base. A second board at least 5 inches wide is attached to the cove base at a 45-degree angle to deflect the light upward and out, away from the wall and into the room. Two other arrangements for homemade ceiling lighting are shown below.

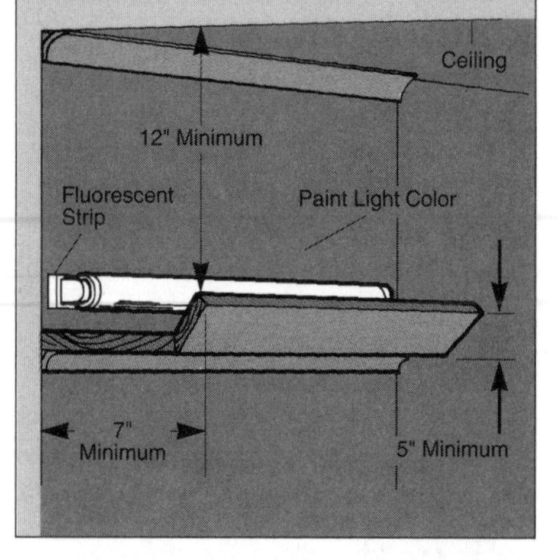

Light Box Between Beams

Use 1x6s and a diffuser panel to create a light box in a beamed ceiling. Paint the inside white and install polished reflectors.

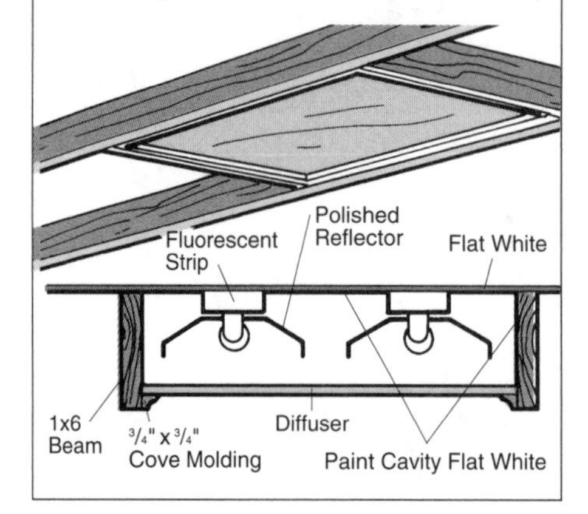

Light in a False Beam

A fluorescent tube hidden in a flash beam uses cove moldings to hold a diffuser panel in place. The beam will throw light downward.

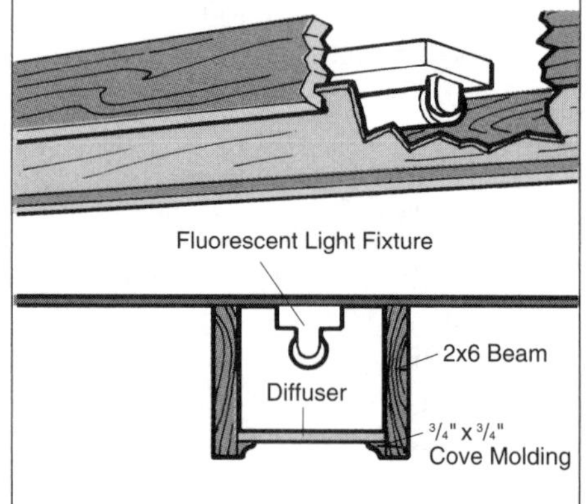

Installing a Junction Box from Below

Where there are finished rooms above a ceiling you must install fixtures from below. Directions for both wallboard and plaster ceilings are given here.

Opening a Wallboard Ceiling

Drill a 1/2-inch hole into the ceiling where you want the fixture; if you hit a joist at this spot, move it a few inches and drill again. Probe through the joists on either side and mark the ceiling along the centers of the joists so that you will be able to nail a patch to the joists. Mark a cut at least 16 inches wide.

Opening a Plaster Ceiling

Use a hammer and cold chisel to open the ceiling enough to expose a strip of lath running perpendicular to the joists. If you see nails in the lath, you are directly under a joist; an absence of nails indicates that you are between joists. Cut a narrow opening along the piece of lath until it extends slightly beyond the outside edges of a pair of joists. Use a keyhole saw to remove the strip of lath directly above the opening.

Installing an Offset Hanger Bar

Clear away any nails or plaster from the bottom of the joists and use wood screws to attach an offset hanger bar as shown at right.

Patching the Wallboard

Cut a patch for the ceiling. Position, the junction box, measure the location, and transfer the outline of the box to the wallboard patch. Cut an opening in the patch for the box. Install the box and nail the patch in place. Finish with tape and joint compound.

Patching Plaster

Follow the directions given on pages 8-9 to repair the gap cut in the lath and plaster ceiling.

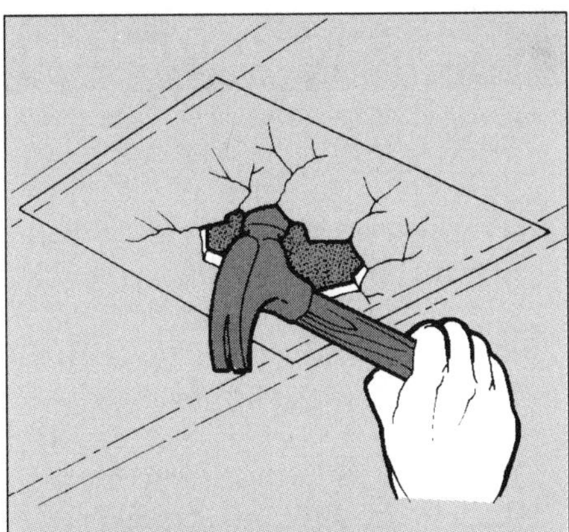

Opening a Wallboard Ceiling. Find an unobstructed point at which to install the box and mark the ceiling for cutting wallboard between joists.

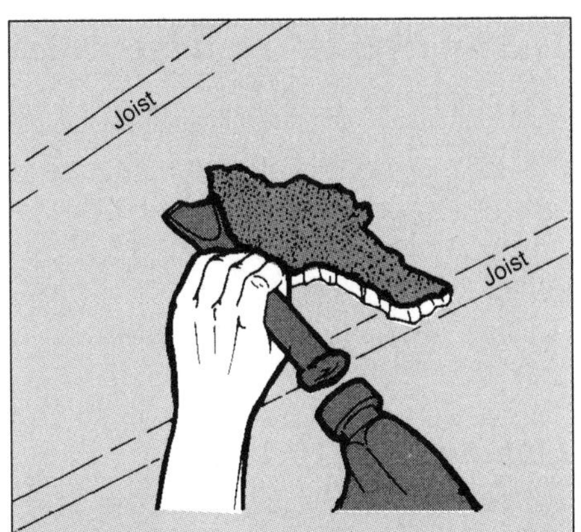

Opening a Plaster Ceiling. Break into the ceiling to find a strip of lath and follow it to the two joists on either side. Protect ceiling along cut by marking with tape.

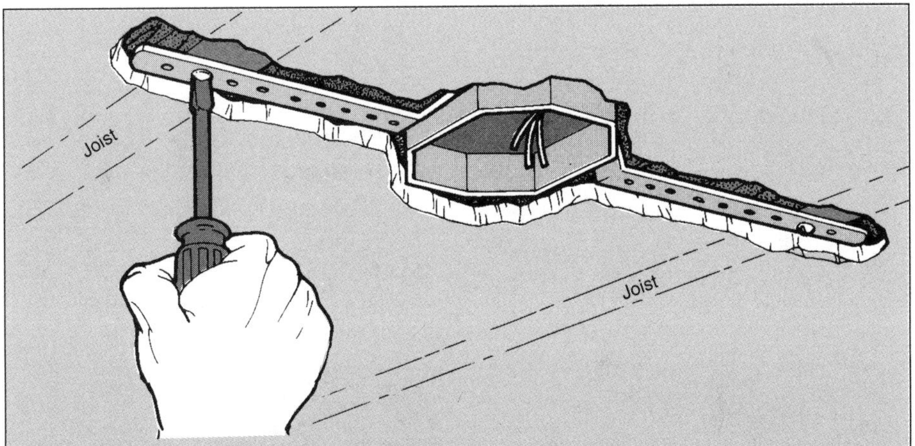

Installing an Offset Hanger Bar. Plaster ceilings do not allow working space between joists; in this kind of ceiling the junction box is attached with an offset hanger bar screwed to the bottoms of the joists. The cut must expose enough of the joists on either side to permit attachment of the bar.

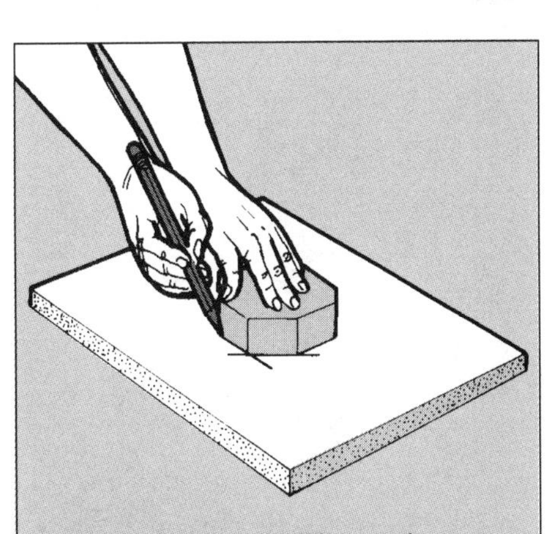

Patching the Wallboard. Cut a wallboard patch to fit and cut an opening in it for the junction box. Attach the box and repair the ceiling with the patch.

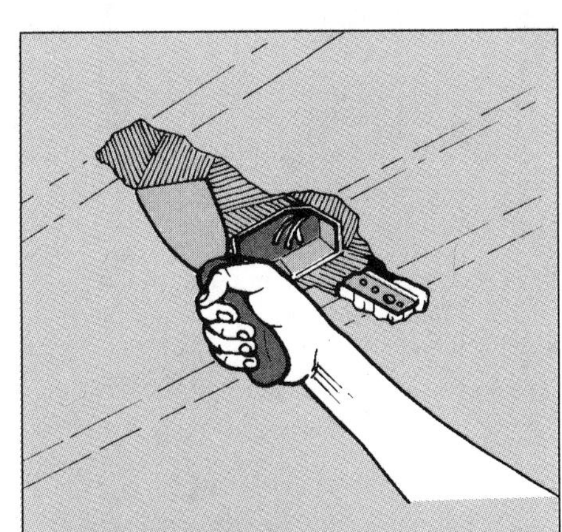

Patching Plaster. Use wire mesh or crumpled, plaster-soaked newspaper to fill the gap; plaster to within 1/16 in. of surface, then finish.

Running Wire Through Finished Ceilings

In finished rooms you must cut into the ceiling and wall in a few places and pull power cable through from a switch or power source with fish tape, an item readily available at hardware and electric supply stores.

Fishing Wire Through Finished Walls & Ceilings

1 Cutting the Wall and Ceiling. Cut the wall and ceiling and drill through the studs between the wall outlet and the point where the cable will run across the ceiling. Protect these studs with metal plates (see page 25). Notch the top plate at the point where the cable will pass, so the ceiling can be closed over it. Push the fish tape up the wall toward the cut at the junction of the ceiling and catch the end with another piece of tape to draw it through. Attach the bottom of the tape to cable and pull it up the wall and out through the ceiling cut.

Fishing Wire to an Unfinished Attic or Crawl Space

From above, drill a 3/4-inch hole through the top plate of the wall (between studs) where the switch or outlet box from which you will run the cable is located. Drop a length of narrow chain through the hole down to the switch or junction box, connect fish tape to the chain, and draw it into the attic. In the attic, attach power cable to the tape and draw it back down to the switch or junction box. Wire to the ceiling box and to the switch or box below.

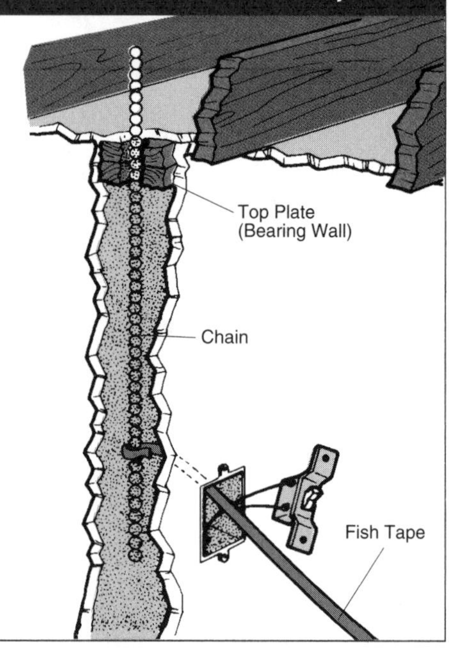

2 Fishing the Cable. Repeat the operation from the junction box to the ceiling cut, pushing tape along the ceiling until you can catch it at the ceiling cut. Attach the tape to the cable and pull it across the ceiling to the junction box. Pull enough cable into the box to do the necessary wiring; attach the cable to the junction box. Wait until all other work is finished before wiring into house circuit.

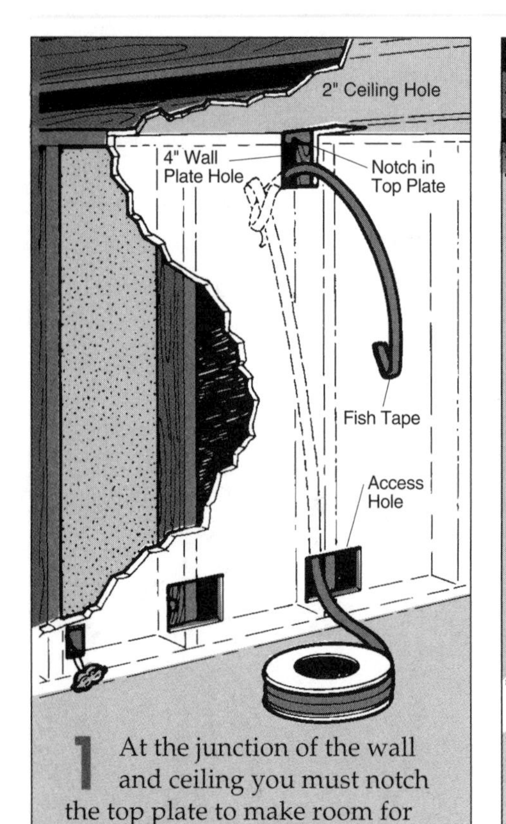

1 At the junction of the wall and ceiling you must notch the top plate to make room for the cable.

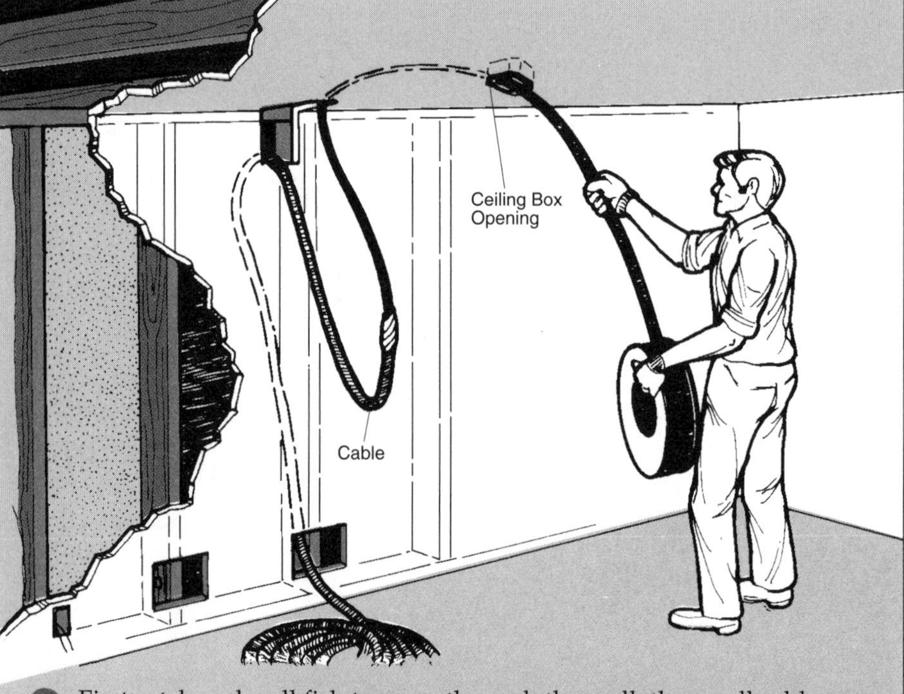

2 First catch and pull fish tape up through the wall; then pull cable through the wall with the fish tape. Push the tape through the ceiling box to the opening at the wall, attach to the end of the cable, and pull it through to the ceiling box.

Installing Track Lighting

There are two basic types of track lighting: one has a line cord that plugs into an outlet, the other is wired directly into a power source. The first type is the easier to install because it does not require bringing power to the track. For the best appearance, plan so that only a minimum amount of line cord runs across the ceiling. Follow manufacturer's instructions for installation. For the type wired directly into a power source, first locate the desired position of the track and follow the procedures given above for installing a junction box in the ceiling.

Caution: *Do all the work with the power in the junction box circuit turned off. Install the track itself using the procedure given below.*

1 Siting the Track. The track will come with an adapter plate to cover the junction box and hold the track connector and electrical housing in place. Assemble these pieces and then splice like-colored wires together from the track's electrical connector to the wiring in the junction box. Attach the adapter assembly to the junction box with the screws provided in the track kit. With the adapter in place, use a straight-edge to draw a line from the center of the mounting slot to the end of the track as a guide for installing the clips that will hold the track to the ceiling.

2 Installing the Clips. Drill holes on the line for anchors into which you will drive the screws that hold the clips to the ceiling. Install the anchors and screw the clips to the ceiling.

3 Installing the Track. Connect the track channel to the electrical connector at the junction box. Make sure that the connection is tight. Slip the track onto the track connector and snap the track into the clips. Tighten the screws on the sides of the clips to fix the track in place. Attach the fixtures according to manufacturer's instructions.

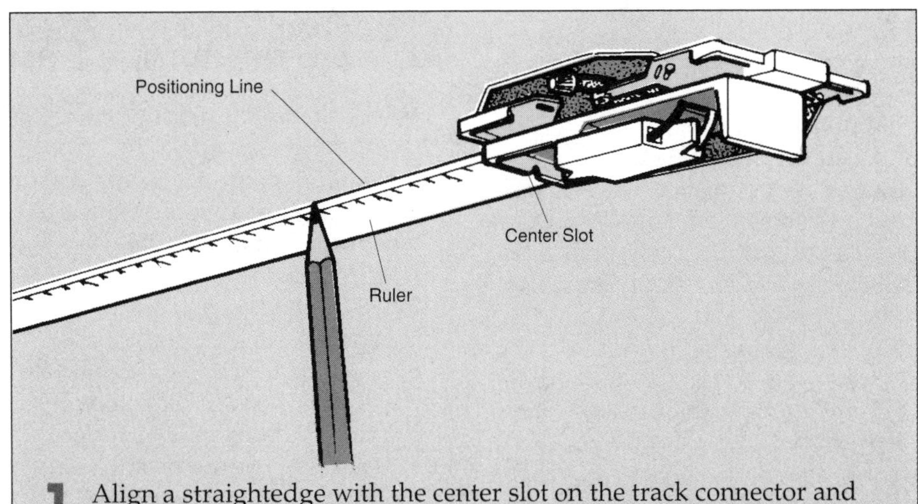

1 Align a straightedge with the center slot on the track connector and draw a line across the ceiling the length of the track. Use a chalk line if the track is long.

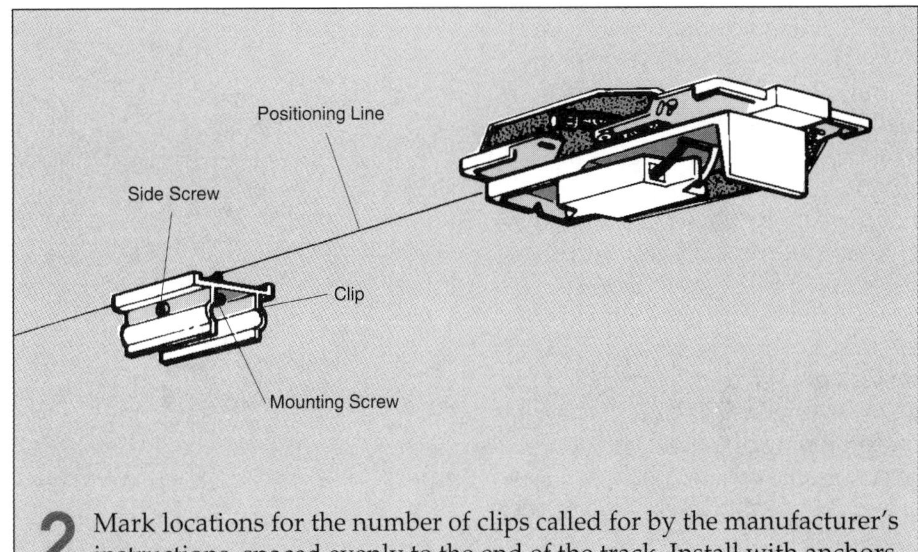

2 Mark locations for the number of clips called for by the manufacturer's instructions, spaced evenly to the end of the track. Install with anchors.

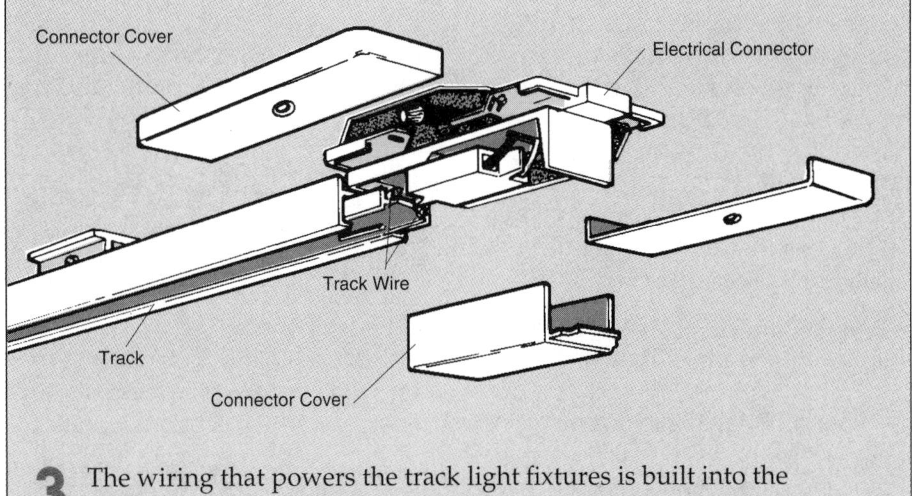

3 The wiring that powers the track light fixtures is built into the track. Push the track firmly into the electrical connector and attach cover plates.

Installing Ceiling Fans

Although it consumes electric power, a ceiling fan can serve as an energy saver. In air-conditioned rooms, the cool air pools on the floor. With the paddles of a slow-turning ceiling fan stirring the air, cool air is distributed throughout the room. Also, moving air seems to make a room cooler. The relatively small amount of power used by the fan achieves the same effect as the much greater amount of power the air conditioner would use if you turned it up in order to cool the room further.

The same is true in heating situations. A room may be comfortable at floor level, but uncomfortably warm near the ceiling—in a sleeping loft or the upper berth of a bunk bed, for instance. A ceiling fan corrects this situation by circulating the rising warm air throughout the room. You need not then burn as much energy to heat the lower part of the room. If you heat with a wood stove, moving the hot air away from the stove with a ceiling fan increases the stove's efficiency by distributing the warmth around the room. It alleviates the problems of a room with an unpleasantly hot zone near the stove and cold air in the rest of the space.

There are a number of features to consider when choosing a ceiling fan: single or variable speed, fixed or moveable paddles, built-in light, one direction or reversible motor (the latter more expensive but very convenient). Installing a ceiling fan is essentially the same as installing a ceiling light fixture, except that the fan must be more securely anchored on the ceiling. Ceilings must be at least 7 feet high.

Wiring for fans is just like that for light fixtures. A fan can be hung from a box where a light fixture was previously hung, to avoid having to install a new box and to fish wire to it, so long as the junction box is secure (reinforced, if necessary).

Flush Mounting

Manufacturer's instructions are specific to each model. The illustrations here show common situations. If you are installing a fan on an unfinished ceiling, or want to take the trouble to cut into a finished ceiling, you can anchor a fan to a joist or header by notching the joist to a depth that brings the edge of the box just flush with the ceiling surface. You also can mount the fan on a block fixed between joists. Follow product instructions—fans of different sizes have different bracing requirements.

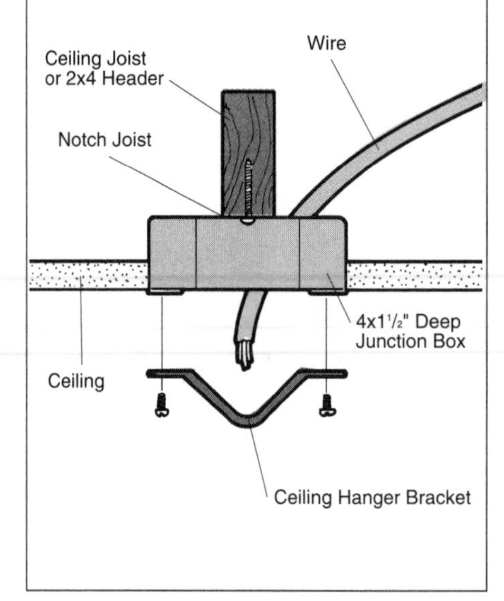

If it is a variable-speed fan you will have to change the switch that controls that box to one with multiple positions.

Check the Local Codes

No matter what kind of fan you install, check to be sure that the installation conforms to local electrical codes. When purchasing a fan, verify that it has been tested and approved according to current electrical standards by a recognized laboratory. Call your local electrical inspector if you have any questions.

Surface Mounting

To mount a fan flush with the surface, find a joist and attach the box to the joist with 2-inch wood screws through the ceiling. Frame around the box to mask it. Run wire across the ceiling to the box according to applicable building code standards.

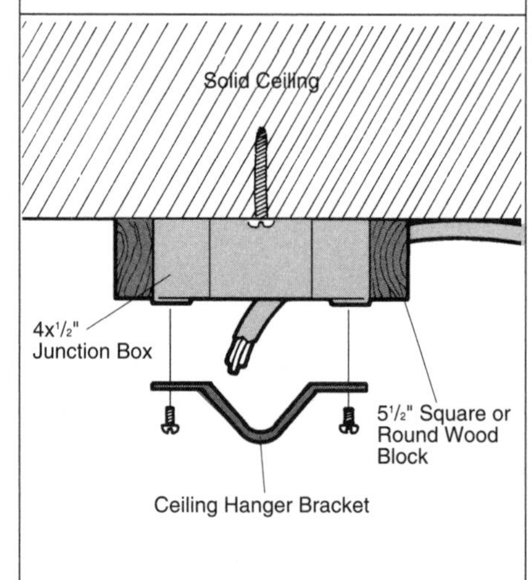

Isolation Mount

The fan will communicate less vibration to the ceiling and run more quietly, if you use an isolation mount in the junction box. The hanger hook fastens over the pins attached to the ceiling.

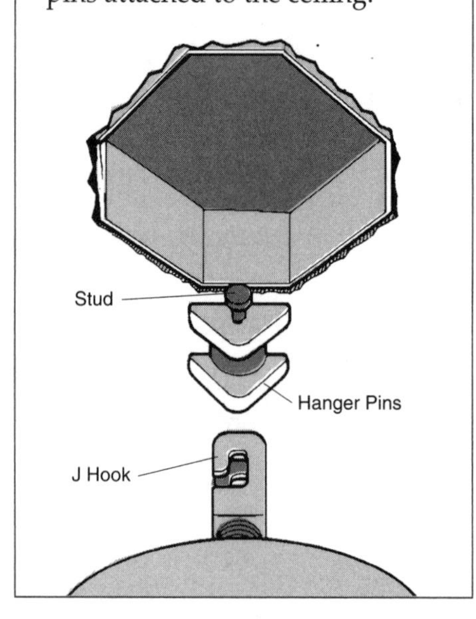

GLOSSARY

Apron The piece of trim around the interior side of a window that sits below the window stool and supports it.

Baseboard A piece of trim, either plain or milled, installed around a room at the base of the walls to conceal the joints of walls and adjoining floor covering.

Baseboard shoe A narrow piece of trim, usually quarter round, attached to the baseboard at the floor to hide any gaps.

Batten Any small strips of lumber used for bracing, for a rigid guideline, for furring, and for some plank-panel walls among many small jobs.

Beam A horizontal structural member that sits on posts or walls and supports the platform or roof above it.

Bearing wall A wall that supports the floors or roof above it. Where there are floors above, the bearing wall often supports the ends of the joists overlapped to span the width of the structure. If a bearing wall is removed, its load-carrying function must be replaced by a beam.

Casing The board lining the inside of a doorway or window frame.

Caulking A soft compound for sealing joints against leaks (of water, air or noise). It may be silicone, neoprene, or one of a variety of other synthetic compounds.

Chair rail A piece of molding that runs around a room about 3½ feet above the floor (approximately the height of a chair back). It evolved from wainscoting.

Cleats Structural members that hold larger parts in place.

Collar beam Beams between rafters that tie them together and support the roof; also beams hung between rafters to support a ceiling.

Concrete A mixture of sand or gravel, Portland cement and water that sets hard.

Corner bead A rigid metal corner attached to an outside wallboard or plaster corner to strengthen it and to provide a foundation for joint compound in finishing the corner.

Countersink Used with a countersink bit, to bore a beveled opening in a pilot hole that is wide enough to accommodate the head of a flathead screw slightly below the surface when fully driven.

Cripple stud The shortened stud above a doorway or window. It maintains the structural integrity of the wall and provides a nailing surface for wallcoverings.

Diffusion panel A translucent piece of plastic under a light fixture or skylight that admits light, but masks what is above.

Double header A header that is double the thickness of the rafters or joists on which it is attached. Used for long spans.

Fish tape Flat, flexible metal tape used for pulling wiring through walls where there is no easy access behind the wall surface.

Flashing Strips of thin metal attached to roof or at the junction of a roof with such structures as a chimney or a skylight, to prevent leaks.

Furring The process of leveling parts of a ceiling, wall, or floor by means of wood strips called *furring strips*, before adding the finish surface.

Hardboard Sheet material composed of wood fibers compressed at high heat. The resulting material is quite hard.

Header A structural member, running at right angles to joists and rafters and connecting to the rest of the framework, at such openings as those for skylights, doorways or stairways.

I-beam A steel beam that in a cross section looks like a capital I. Used to span distances greater than wood beams can support.

Jack stud A stud in a doorway or below a window, that does not reach top plate.

Jamb The vertical inside face of a door or window opening.

Joist A horizontal structural member that supports a floor and a ceiling in the case of middle floors, and a ceiling alone in the case of a top floor. Joists are often tied together with cross bridging for extra rigidity.

Knee wall A wall built under a sloping roof that does not reach to the height of the ceiling.

Lath A strip of wood used as a foundation for plaster walls and ceilings.

Molding Various types of wood used for decorative or practical trim. It is usually applied to conceal joints of dissimilar surfaces.

Nailer A piece of wood added to a framework to provide a nailing surface for all edges of something (usually a sheet of wallboard or paneling) to be attached.

On center A phrase designating the distance between the centers of regularly-spaced holes, or such parts as studs in a wall.

Particleboard Inexpensive sheet material composed of wood chips, or particles, and adhesive pressed into sheets. Fairly brittle and not as easy to nail through as wood.

Partition A wall that divides space but plays no part in the structural integrity of a building.

Plaster A mixture of sand, lime, and water of a consistency that can be applied to a wall or ceiling with a trowel. Various plasters have different uses, such as for patching holes, or for finishing a smooth surface.

Plate A horizontal member, at the top or bottom of a wall, to which the studs are fixed.

Prehung door A door that is delivered from the factory already hung in its casing so that it need only be attached to the doorway.

Rafter A structural member that supports a pitched roof, serving for the roof sheathing the same purpose as joists for floors.

Shims Thin wood wedges used for tightening the fit between pieces, as when they fill the gap between the top plate and joists in new wall construction, or for leveling strips of furring over an uneven surface.

Spacer block A piece of lumber used between larger attached members to maintain a uniform separation (for instance, between two studs at the outside corner of a frame wall). Also called a *filler block.*

Starter hole A hole drilled inside the outline of a shape to be cut out of a sheet of material to permit inserting a saw blade to make a cut.

Stud Vertical member of a frame wall, usually placed at either end and one every 16 inches on center to facilitate covering with standard-width wallboard or paneling. May be 2x3, 2x4, 2x6 depending on the structural requirements of the wall.

Tongue and groove The milling treatment of the edges of a board resulting in a protruding tongue on one side and groove the same size on the other. For the purpose of joining several boards.

Wainscot Paneling that reaches from the floor to about one third the height of a wall, capped with molding.

Wallboard Also known as gypsum board, drywall, and plasterboard; a paper-covered sandwich of gypsum plaster used as the primary wall covering in almost all homes.

Window stool Molding that connects the window with the interior of a room, often referred to as the window sill.

INDEX